Jens Köhler

Tunable Security for Deployable Data Outsourcing

Tunable Security for Deployable Data Outsourcing

by

Jens Köhler

Scientific Publishing

Dissertation, Karlsruher Institut für Technologie (KIT)
Fakultät für Informatik, 2015

Tag der mündlichen Prüfung: 9. Juni 2015
Referenten: Prof. Dr. Hannes Hartenstein, Prof. Dr. Alexander Pretschner

Impressum

KIT Scientific Publishing

Karlsruher Institut für Technologie (KIT)
KIT Scientific Publishing
Straße am Forum 2
D-76131 Karlsruhe

KIT Scientific Publishing is a registered trademark of Karlsruhe Institute of Technology. Reprint using the book cover is not allowed.

www.ksp.kit.edu

Print on Demand 2015

ISBN 978-3-7315-0411-5
DOI 10.5445/KSP/1000048084

Tunable Security for Deployable Data Outsourcing

zur Erlangung des akademischen Grades eines

Doktors der Ingenieurwissenschaften

von der Fakultät für Informatik
des Karlsruher Instituts für Technologie (KIT)

genehmigte

Dissertation

von

Jens Köhler

aus Witzenhausen

Tag der mündlichen Prüfung: 09.06.2015

Erster Gutachter: Prof. Dr. Hannes Hartenstein
Karlsruher Institut für Technology (KIT)

Zweiter Gutachter: Prof. Dr. Alexander Pretschner
Technische Universität München (TUM)

Zusammenfassung

Der Trend in der heutigen vernetzten Welt geht dahin, IT-Dienste von Drittanbietern zu nutzen. Sowohl für Endnutzer als auch Dienstbetreiber existiert eine Vielfalt an Angeboten, die eine Auslagerung von Daten an Drittanbieter erfordern. Beispielsweise verspricht die *Auslagerung von Datenbanken* zu externen Cloud-Anbietern Kostenersparnisse für Dienstbetreiber. Darüber hinaus können sich Dienstbetreiber mit Hilfe von Technologien des *föderativen Identitätsmanagements* auf digitale Identitäten verlassen, welche die Endnutzer bereits bei anderen Parteien etabliert haben und so vermeiden, selbst digitale Identitäten verwalten zu müssen. Um auf Dienste zugreifen zu können, müssen Nutzer eine Vielzahl an Anmeldedaten verwalten. Um Anmeldedaten zu speichern, greifen sie zunehmend auf *Credential-Repositories* wie etwa Passwort-Manager zurück. Angebote wie die genannten implizieren, dass Daten zu mehreren verschiedenen Parteien ausgelagert werden, die teilweise verschiedenen Rechtsprechungen unterliegen und individuellen Richtlinien bezüglich dem Schutz sowie der Nutzung der ausgelagerten Daten folgen. Vor diesem Hintergrund stellt die Durchsetzung von klassischen Schutzzielen wie Datenvertraulichkeit, -integrität und -verfügbarkeit eine Herausforderung dar.

Mittels Sicherheitsmechanismen wie beispielsweise Verschlüsselung ist es möglich, Schutzziele durchzusetzen, bevor die Daten ausgelagert werden. Allerdings beeinträchtigen Sicherheitsmechanismen oft andere Qualitätseigenschaften wie etwa Effizienz negativ. Unnötige negative Effekte auf Qualitätseigenschafen können vermieden werden, indem Ansätze auf die Anforderungen des gegebenen Einsatzszenarios maßgeschneidert werden und nur Sicherheitsmechanismen eingesetzt werden, die wirklich nötig sind. Allerdings sind solche maßgeschneiderten Ansätze oft nicht in Einsatzszenarien mit anderen Anforderungen einsetzbar. Oftmals muss ein maßgeschneiderter Ansatz neu entworfen und implementiert werden, um in einem Einsatzszenario einsetzbar zu sein, für das er nicht konzipiert war. Ansätze die es erlauben, die Abwägungen zwischen Sicherheits- und anderen Qualitätseigenschaften *nach* der Entwurfs- und Implementierungsphase einzustellen, können in einer größeren Menge an Einsatzszenarien eingesetzt werden und mit Anforderungen umgehen, die sich über die Zeit weiterentwickeln.

In dieser Dissertation wird untersucht wie a) eine passende Kombination von Sicherheitsmechanismen ermittelt werden kann, um einen Ansatz auf die Anforderungen eines speziellen Einsatzszenarios maßzuschneidern und b) der Prozess zu Findung einer passenden Sicherheitsmechanismenkombination automatisiert werden kann. Basierend auf den resultierenden Erkenntnissen können Ansätze geschaffen werden, deren Eigenschaften ohne Neukonzipierungs- und Neuimplementierungsaufwand passend auf Einsatzszenarioanforderungen eingestellt werden können. Somit adressiert diese Dissertation die folgende fundamentale Forschungsfrage: *Wie können Sicherheitseigenschaften einstellbar gemacht werden, um breit einsatzbare Datenauslagerungsansätze zu ermöglichen?* Diese Dissertation zeigt auf, wie Errungenschaften aus den Gebieten Policy-based Management und Operations Research mit Sicherheitsmechanismen zu einer Methodik für die Erstellung von einstellbaren Ansätzen kombiniert werden können. Die vorgeschlagene Methodik wird angewendet um anerkannte wissenschaftliche Datenauslagerungsprobleme in den Gebieten 1) Datenbankauslagerung, 2) föderatives Identitätsmanagement und 3) Credential-Repository Auslagerung zu adressieren:

1) Die Dissertation enthält eine Studie bezüglich existierenden Sicherheitsmechanismen, welche die Vertraulichkeit von Datenbanken durchsetzen und gleichzeitig die Ausführung von Anfragen auf die Datenbank ermöglichen. Weiterhin werden Datenbankauslagerungsansätze mit einstellbaren Vertraulichkeits- und Anonymitätsgarantien vorgestellt, welche die Effizienz hinsichtlich Anfragelatenz, Speichernutzung und Netzwerkkapazität maximieren soweit es die Sicherheitsanforderungen zulassen. Insbesondere wird der Securus-Ansatz vorgestellt, der es erlaubt, Vertraulichkeitsanforderungen, angenommene Angreifer und Anfragbarkeitsanforderungen zu spezifizieren. Falls spezifizierte Anforderungen konfligieren, ermittelt Securus diese und präsentiert sie dem Nutzer um ihn bei der Konfliktauflösung zu unterstützen. Für konfliktfreie Anforderungen ermittelt Securus eine effizienz-optimierte Kombination von Sicherheitsmechanismen, die die Anforderungen erfüllen, und implementiert diese. Die in der Arbeit durchgeführte Leistungsbewertung zeigt, dass Securus eine maßgeblich höhere Leistung erreichen kann als Vertraulichkeits-durchsetzende Datenbankauslagerungsansätze bei denen Vertraulichkeitseigenschaften nicht einstellbar sind. Des Weiteren wird der Dividat Ansatz präsentiert, der zur Erstellung von Leistungsmodellen für anonymisierte Datenbanken genutzt werden kann. Derartige Leistungsmodelle sind nötig, um die Optimierung von Indizierungsstrategien von anonymisierten Daten zu automatisieren und so die Effizienz von Anfragen unter Einhaltung von Anonymitätsanforderungen zu steigern.

2) Dienste, die auf föderativen Identitätsmanagementansätzen basieren, sind für einen Endnutzer nicht verfügbar, wenn die Heimatorganisation des Nutzers nicht verfügbar ist und somit keine Identitätsdaten liefern kann. Die Dissertation stellt den Occasio-Ansatz vor, mit dem Heimatorganisationen Nutzeridentitäten sicher zu einer hoch-verfügbaren, externen Partei auslagern können, der nicht zugetraut wird die Vertraulichkeit und Integrität der Identitätsdaten sicherzustellen. Insbesondere erlaubt es Occasio, die nötige Abwägung zwischen Aktualität und Verfügbarkeit von ausgelagerter Identitätsinformation an die Anforderungen des Einsatzszenarios anzupassen.

3) Credential-Repositories sind anfällig für Passwort-Rate-Angriffe, falls ein Angreifer das System, auf dem das Credential-Repository betrieben wird kompromittieren kann und die enthaltenen Geheimnisse nur mit einem schwachen Passwort geschützt sind. Die Dissertation stellt den Credis-Ansatz vor, der ein auf mehrere Parteien verteiltes Credential-Repository basierend auf Secret-Sharing Ansätzen erstellt. Ein verteiltes Credential-Repository ermöglicht es, sicher kryptographisch schwache Passwörter gegen Geheimnisse wie kryptographische Schlüssel zu tauschen, sofern nicht alle Parteien kompromittiert sind. Credis optimiert die durch die Secret-Sharing Ansätze nötige Abwägung zwischen Vertraulichkeit und Verfügbarkeit der Geheimnisse automatisch hinsichtlich der Einsatzszenarioanforderungen.

Abstract

In today's networked world the trend to make use of IT services that are provided by external parties accelerates. A variety of offerings that require to outsource data to third parties have emerged both for end users and service providers. For instance, service providers have the option to cut costs by *outsourcing databases* to various external cloud providers. Furthermore, service providers can avoid having to manage digital identities that are needed to authenticate and authorize end users by relying on identities that the end users have already established with other parties based on *federated identity management* technologies. To access services, users have to remember a lot of credentials, a task for which they increasingly rely on *credential repositories* such as password managers that allow to store credentials. Such data outsourcing options imply to outsource data to multiple different parties that are subject to different jurisdictions and have individual policies on how to protect and use the outsourced data. Thus, enforcing traditional security characteristics such as data confidentiality, integrity, and availability constitutes a challenge in such a setting.

It is possible to enforce the security characteristics before outsourcing the data by security mechanisms like encryption. However, in many cases these security mechanisms negatively affect other quality characteristics like efficiency. Unnecessary negative effects on quality characteristics can be avoided by tailoring an approach to apply only security mechanisms that are really needed to satisfy the security requirements of the given scenario. However, such an approach that is tailored to satisfy the requirements of a single scenario is often undeployable in scenarios with different requirements. This limits the deployment potential of the approach's implementation. Furthermore, even in the scenario for which the approach was tailored, applying the approach can become inadequate if the scenario's requirements evolve over time. To deploy a tailored approach in a scenario with different requirements than the original scenario, the approach has to be re-designed and re-implemented in many cases. Approaches that allow to tune security trade-offs *after* the design and implementation phase can be used both to cope with evolving scenario requirements and to cover a larger application area.

In this thesis we explore a) how a suitable combination of security mechanisms can be determined when designing an approach that is tailored to satisfy specific deployment scenario requirements and b) how to automate the process of finding a suitable security mechanism combination. Based on our insights, approaches can be built that can be to tuned to satisfy deployment scenario requirements without re-design or re-implementation effort. Thus, we address the following fundamental research question: *How can security characteristics be made tunable to enable deployable data outsourcing approaches?*

We address this question by proposing a methodology on how to build tunable approaches that combines findings from the domains of policy-based management and operations research with security enforcement mechanisms. Furthermore, we apply the proposed methodology to make contributions that address scientifically acknowledged data outsourcing problems in the domains of 1) database outsourcing, 2) federated identity management, and 3) credential repository outsourcing:

1) To address the problem of securely outsourcing databases, we survey existing security mechanisms that preserve database confidentiality and allow to execute database queries on the outsourced data at the same time. Furthermore, we propose approaches that can be used to tune the confidentiality and anonymity of outsourced databases in order to improve efficiency with regard to query latency, storage capacity, and transmission overhead. We present Securus, an approach that allows to specify confidentiality requirements, against which attackers the outsourced data has to be protected, and how the data will be queried. In case the specified requirements are conflicting, Securus narrows down which requirements are conflicting and presents them to the user to aid in conflict resolution. If the requirements are not conflicting, Securus automatically generates a software adapter that incorporates security mechanisms to satisfy the confidentiality requirements and is efficiency optimized for the specified queries. We measured the query performance of Securus and found it to be significantly higher than that of other confidentiality-enforcing database outsourcing approaches that cannot be tuned. Furthermore, we propose Dividat, an approach to build performance models for anonymized databases. Building performance models for anonymized database indexes constitutes a step towards automating the optimization of indexing strategies for anonymized databases and tuning anonymity for increased efficiency.

2) One problem of federated identity management approaches is that services are not available to end users if the home organization that stores the user's digital identity is unavailable. We contribute Occasio, a tunable approach that allows home organizations to securely outsource user identities to highly available external parties that are not trusted to enforce appropriately security characteristics such as the confidentiality and the integrity of identity data. Among others, Occasio allows to tune the inherent trade-off between integrity and availability of the outsourced identity information.

3) Credential repositories are often vulnerable to password guessing attacks if an attacker compromises the system that hosts the credential repository and a weak password is used to protect the stored secrets. We contribute Credis, a tunable approach to build credential repositories that are distributed on multiple parties based on secret sharing schemes. Based on a distributed credential repository, it is possible to securely trade cryptographically weak secrets such as passwords for strong secrets such as cryptographic keys. Credis allows to fine-tune the trade-off between secret confidentiality and secret availability that is inherent to all secret sharing schemes.

Contents

List of Figures

List of Tables

1 Introduction

"Everything should be made as secure as necessary, but not securer."

Ravi Sandhu - Executive Director and Chief Scientist
Institute for Cyber Security (ICS)
University of Texas at San Antonio

In today's networked world the trend to use IT services that are provided by external parties accelerates. A variety of IT service offerings have emerged both for end users and service providers that imply to outsource data to external parties. Service providers have the option to cut costs by *outsourcing databases* to various external cloud providers [AFG+09]. Furthermore, service providers can avoid having to manage digital identities that are needed to authenticate and authorize end users by relying on identities that the end users already have established with other parties based on *federated identity management* technologies [CI09]. To access services, users have to remember of a lot of credentials including passwords and cryptographic private keys, a task which they increasingly outsource to *credential repositories* such as password managers that allow to store credentials.

In this section we outline the benefits as well as the security challenges that arise when outsourcing data in the domains of database outsourcing, federated identity management, and credential repositories. Furthermore, we highlight security trade-off situations that occur in data outsourcing settings and motivate the need for approaches with tunable security properties. Based on these motivations we state the research questions that are addressed in this thesis and provide an overview of the contributions as well as the structure of the thesis.

1.1 Benefits of Data Outsourcing

Outsourcing data offers many benefits in various domains. In this thesis we focus on 1) outsourcing databases, 2) relying on outsourced digital identities, and 3) outsourcing user secrets to so-called credential repositories. Some of the benefits that come with data outsourcing in each of these domains are illustrated in the following.

1) Database outsourcing: From the perspective of a service provider (SP), outsourcing databases to external cloud storage providers (CSPs) can yield several benefits [AFG+09]. One benefit is *scalability* due to the large resource pool of the cloud provider, i.e., the ability to scale the capacity of the database and the query workload. Other aspects are *cost benefits* that result from a high *elasticity*, i.e., the ability to scale the capacity of the database down on demand, and the *pay-as-you-go business model* of common cloud providers that only requires to pay for resources that are requested. This also implies that, from the service providers perspective, no big investments have to be made up-front to get access to a large pool of resources. Another benefit is that *no know-how has to be maintained by the service provider* on how to operate and manage highly available infrastructure as the cloud provider operates the infrastructure and offers a highly available platform to its customers. For instance, developers of applications that are run on the GoogleAppEngine cloud infrastructure[1] do not have to worry about the availability of the underlying hardware or network outages.

2) Identity outsourcing: In order to enable service providers (SPs) to authenticate users and make authorization decisions, digital identities of these end users have to be managed. Based on federated identity management approaches, the task of managing and maintaining the digital identities of end users can be outsourced by relying on digital identities that the user has already established with other organizations, so-called identity providers (IdPs). Besides *relieving the service provider* of having to keep identities and the according authorization tokens up-to-date, using federated identity management approaches can also lead to a *better user experience*, as existing accounts can be used to access new services without having to remember any additional credentials. This in turn leads to *fewer password reuse* at multiple services and therefore reduces the risk of user passwords becoming compromised.

3) Credential outsourcing: Whereas federated identity management simplifies the credential management for users, it is unrealistic to believe that users will only have to manage credentials for a single digital identity in the future. For instance, banks will most likely not allow users to authenticate based on their Facebook identity alone and users want keep some of their digital identities separated for privacy reasons. Thus, users still have to manage a variety of credentials. Ideally, they should use unique passwords for each of their user accounts. Furthermore, some users have to manage user certificates for applications such as encrypted or signed email communication. In general, if a user encrypts data she has to manage cryptographically strong encryption keys. While memorizing multiple passwords is cumbersome for most users, memorizing cryptographically strong encryption keys is more or less impossible. The user can rely on credential repositories to store and retrieve creden-

[1] https://appengine.google.com/ [last visited on March 2015]

tials. In order to be able to retrieve credentials on arbitrary end-devices the credential repository has to be highly available. To enhance availability and minimize the effort to operate the credential repository, the user can outsource the credential repository to a highly available third party.

1.2 The Need for Tunable Security

One major challenge of outsourcing data is that the ability to control who can access the data is forfeited to the party that stores the outsourced data. A risk manager may decide that an approach that outsources data to third parties constitutes too much of a risk due to a high probability that certain protection objectives are undermined by relying on the third party. In particular, such protection objectives include security characteristics like data confidentiality and data integrity. One potential risk factor is that the third party may be malicious and deliberately undermines protection objectives for its own gain, e.g., by selling the outsourced data. Even if the third party is honest, its employees (e.g., admins) might not be or the third party may be forced to cooperate with governmental agencies for legal reasons [Nie06]. The third party can also be compromised by external attackers if its infrastructure is not sufficiently protected.

In some cases these risks can be mitigated by contracts that regulate accountability, by certifications of the third parties according to common security standards such as the ISO 270xx standard series [ISO14b, ISO14c], by relying on parties that are not subject to jurisdictions that force them to undermine protection objectives, or by relying on "commonly trusted" third parties that have a good reputation for enforcing protection objectives as intended. In this thesis, we focus on another approach: to enforce security characteristics before outsourcing the data by technical security mechanisms like encryption. While security mechanisms enforce security characteristics, they can negatively affect other quality characteristics. If the problem setting inherently contains one of these trade-offs between security properties and other quality characteristics, designing and implementing approaches that satisfy the requirements of all possible deployment scenarios is impossible. To satisfy the requirements of a specific scenario, approaches often have to be tailored to the specific scenario in which they will be deployed. In the following, we show how security mechanisms that enforce security characteristics can affect other quality characteristics such as efficiency, usability, and availability in the problem settings of database, identity, and credential outsourcing. Based on these observations we then motivate the need for approaches with tunable security properties.

1) Database outsourcing: When outsourcing databases to cloud providers, the risk of data disclosure can be mitigated by encrypting the entire database before outsourcing it. However, this induces efficiency overheads for evaluating queries on the outsourced database. For instance, it is no longer possible to retrieve only records that match certain criteria, as the cloud provider cannot evaluate which encrypted records match the criteria. Thus, to evaluate a query on the database, the whole encrypted database has to be downloaded from the cloud provider and then be decrypted. This process induces a high efficiency overhead in terms of query latency and network transmission. To optimize efficiency, an approach has to protect only the parts of the database that require protection, i.e., tune its security properties to the requirements of the given scenario.

2) Identity outsourcing: To make an authorization decision, a service provider often requires authorization tokens that are part of the digital identity of the user who accesses the services. For instance, such an authorization token can be a membership in a group. To make sure that the authorization decision is correct, the service provider has to make sure that the authorization tokens are up-to-date which is not trivial in the federated identity management setting since the authoritative source of the authorization tokens is the identity provider rather than the service provider. The risk of making incorrect authorization decisions can be mitigated by letting the service provider query up-to-date authorization tokens from the identity provider each time they are needed. However, this affects the availability of the service. If the identity provider is not available to provide authorization tokens, the service provider cannot be sure to make a correct authorization decision and has to decline access to the user. To maximize the availability of service providers, an approach has to be able to tune the time span in which outdated authorization tokens are still accepted to the requirements of the given scenario.

3) Credential outsourcing: When a user outsources a credential repository to a third party in order to make the contained credentials highly available and easily accessible, the user faces the risk that the third party is compromised and the credentials are revealed to the attacker. To mitigate the risk of a compromised user credential repository, the credential repository can be split up on multiple parties so that a single compromised party is not able to reveal any credentials. However, splitting the credential repository up on multiple parties affects its availability, since all parties have to be available for the user to be able to retrieve credentials. Ideally, such a credential outsourcing approach should allow to tune its security properties by splitting up the credential repository in such a way that the risk of compromised secrets is mitigated to an acceptable level and the credential repository is as available as possible.

The examples for database, identity, and credential outsourcing illustrate that enforcing security properties can negatively affect other quality properties such as efficiency and availability. Thus, many IT systems are forced to choose between enforcing certain security properties and achieving other quality properties [San03, EY07]. By making this decision, the application area of these systems is limited as they can only be deployed in scenarios with matching requirements. For instance, an approach that encrypts databases before outsourcing them cannot be used in scenarios in which a high efficiency is required. Likewise, an approach that outsources a database in plaintext to guarantee a high efficiency cannot be applied in scenarios in which confidentiality is an imperative requirement. This application area limitation is particularly problematic if it can be assumed that the requirements of the deployment scenario change over time, possibly even during the implementation phase of the approach [Che03].

Based on these observations, we argue that a paradigm shift is necessary when designing approaches for problem settings that contain inherent trade-offs between security and other quality characteristics. Standard approaches that have fixed security properties are not suited for a problem setting with diverse scenario requirements, as such approaches would have to be constantly re-designed and re-implemented to satisfy requirements of different scenarios or to cope with scenario requirements that change over time. Thus, tunable approaches are required that allow to adapt their security properties to the individual scenario requirements after the design and implementation phase.

1.3 Problem Statement and Research Questions

To keep the application area of an approach wide, the approach has to be *tunable*, i.e., it has to be possible to parameterize the approach to satisfy the requirements of the scenario in which it is to be deployed. In comparison to static approaches, tunable approaches do not have to be re-designed and re-implemented to alter trade-offs between security and quality properties. We advocate the view that tunable approaches should not consider the trade-offs between security and other quality characteristics as a binary one but should enable to tune the trade-offs in a fine-grained manner. This maximizes the area of application as many scenario requirements since security and other quality characteristics are not binary either. To achieve and fine-tune trade-offs between security and other quality characteristics we address the following fundamental research question in this thesis:

How can security characteristics be made tunable to enable deployable data outsourcing approaches?

To approach this generic question, we show how tunable approaches can be built that address scientific challenges in the domains of *database outsourcing* (Section 1.3.1), *federated identity management* (Section 1.3.2) and *credential repositories* (Section 1.3.3). Based on these contributions, we show how fine-grained trade-offs between security characteristics and other characteristics can be achieved and tuned to enable deployable approaches.

1.3.1 Database Outsourcing

In the Database-as-a-Services (DaaS) problem setting, security properties like data confidentiality or the anonymity of individuals have to be enforced before outsourcing the database to the external, untrusted cloud provider. As we showed in Section 1.2, security mechanisms that enforce security properties like confidentiality negatively affect efficiency and a trade-off situation exists between efficiency and security properties of database outsourcing approaches. Efficiency can be improved by only protecting the data that requires protection. The individual confidentiality requirements of each scenario in which data is outsourced are highly diverse. For instance, entirely different parts of the data have to be protected when outsourcing either research data or patient records. This underlines the necessity for a tunable approach that satisfies the security requirements of a scenario and maximizes efficiency as much as possible. The research question that we address in this thesis with regard to secure database outsourcing is the following:

How can database outsourcing approaches tune their security properties to satisfy security and efficiency requirements?

In particular, we investigate how efficiency with regard to performance and resource utilization can be maximized by tuning the confidentiality and anonymity guarantees of approaches to satisfy individual scenario requirements.

1.3.2 Identity Outsourcing

In the federated identity management setting, the availability of the service providers depends on the availability of the identity provider that authenticates the user and provides the service provider with information that is needed to make access control decisions. If the identity provider is not available to authenticate a user that wants to access a service, the service is not available to the user. Thus, identity providers are typically obligated to be highly available. Especially for small-scale home organizations that do not operate computing centers, operating highly available identity providers constitutes a challenge. As we show in this thesis, securely outsourcing identity providers to highly available cloud providers is possible, but requires the technical enforcement of certain security properties if the cloud provider is not trusted to enforce them. In particular, the freshness of the outsourced identities has to be enforced to allow authorization decisions based on up-to-date identity information. However, applying security mechanisms that enforce the security properties have a negative effect on other quality properties. Whether security mechanisms to enforce security properties have to be applied depends strongly on the individual requirements of the given scenario. We address the following research question with regard to outsourcing identities to highly available cloud providers:

How can approaches to outsource identity providers tune security characteristics to satisfy security, usability, availability, and efficiency requirements?

1.3.3 Credential Outsourcing

Modern users have to manage a multitude of credentials and use them on various end-devices. Credential repositories can help users to achieve this. By outsourcing credential repositories to highly available parties or devices, the credential repository can be accessed from arbitrary end-devices. However, the confidentiality of the contained credentials has to be enforced by security mechanisms if the parties that host the credential repository are not trusted to enforce confidentiality appropriately. Enforcing confidentiality via encryption requires the user to memorize a cryptographically strong encryption key as user passwords typically do not contain enough entropy to be applied securely for encryption. An alternative to encryption are secret sharing schemes which split the credentials up on multiple parties, forming a *distributed credential repository*. Such schemes guarantee confidentiality as long as not all of the parties are compromised and allow to access secrets via a memorable user generated password. However, the credentials can only be accessed when all parties are available. Thus, a trade-off between the confidentiality and availability of a distributed credential repository exists that has to be adapted to the individual scenario requirements. We address the following research question with regard to distributed credential repositories:

How can the confidentiality of credentials be tuned in distributed credential repositories to satisfy confidentiality and availability requirements?

1.4 Main Contributions of this Thesis

In the following we provide an overview of the contributions presented in this thesis.

Based on the ISO 250xx standards [ISO14a] on software and system quality we provide a **conceptual framework for tunable security**. We show how the concept of tunability relates to other research areas and perspectives including risk management and secure system development. Based on the conceptual framework we propose a **methodology to build tunable approaches** that allow to tune security trade-offs in a fine-grained manner. In particular, we show how the methodology can be used to beneficially combine methods from operations research, policy based management, and secure system development to build approaches with tunable security properties. We apply the methodology to address acknowledged research challenges in the domains of database outsourcing, federated identity management, and distributed credential repositories. The resulting tunable approaches constitute substantial contributions on their own. The domain-specific contributions are summarized in the following.

Database Outsourcing

- **Taxonomy and assessment of security mechanisms that enforce confidentiality in databases** [KJH15]. We propose a taxonomy that can be used to assess existing security mechanisms to preserve data confidentiality in databases while still being able to efficiently evaluate specific database queries. Such security mechanisms are denoted as confidentiality-preserving indexing approaches (CPIs) in the following. We apply this taxonomy in a comprehensive survey of CPIs to categorize them according to the provided level of security, their robustness against various attacker models and their ability to evaluate queries efficiently.

- **Securus** [JKH12, KJ14, KJH14], an approach to find an efficiency-optimized combination of CPIs that satisfies user-specified requirements with regards data confidentiality requirements, against which attackers the outsourced data has to be protected, and how the data will be queried. In case no such CPI combination exists, some of the specified requirements are not satisfiable together and Securus presents the conflicting requirements to the user. Securus tunes confidentiality in a fine-grained manner by considering the CPIs' differences with regard to confidentiality guarantees against various attacker models. Securus automatically generates a software component that implements the CPI approaches which enforce the specified deployment scenario requirements. The user can transparently use this software component to outsource data and execute queries without being required to adapt the design or implementation of Securus to the scenario. Our query performance measurements show that by being able to tune the confidentiality properties of the outsourced database to satisfy the requirements, Securus can significantly outperform other secure database outsourcing approaches that cannot be tuned.

- **Dividat** [KH14], an approach to build efficiency models for outsourced, anonymized databases. Based on the example of ℓ-diversified databases we show that it is not trivial to find an efficiency optimal strategy to execute queries on anonymized data which makes performance models necessary. We exemplarily apply Dividat to build performance models for ℓ-diversified databases and show how they can be used to optimize the efficiency of query execution. The efficiency models can be used by approaches to tune anonymity for efficiency, i.e., by relaxing anonymity requirements a more efficient query execution strategy can be found.

Identity Outsourcing

- **Occasio** [KH13], a tunable approach to securely outsource identity providers to highly available external providers that cannot be trusted to adequately enforce security characteristics like confidentiality and integrity. We propose security mechanisms that can be used to enforce security against malicious external providers who try to read confidential identity data and manipulate authorization decisions. Additionally we investigate how the security mechanisms affect other quality characteristics and show that Occasio's security properties can be tuned to satisfy deployment scenario requirements in order to omit these negative effects whenever the scenario requirements allow for it.

Credential Outsourcing

- **Credis** [KMH13], a tunable approach to build distributed credential repositories that store credentials and allow to retrieve the credentials by providing a weak, user-generated password. Credis protects the confidentiality of the outsourced credentials by making use of existing secret sharing schemes that split up the credentials on multiple parties. By naively applying these schemes, all parties have to be compromised to undermine the credentials' confidentiality, yet all parties have to be available to allow a legitimate user to retrieve the credentials. Credis applies the schemes in a more sophisticated way that allows to tune the trade-off between confidentiality and availability in a more fine-grained manner. Furthermore, Credis automatically determines a strategy to apply the schemes that satisfies specified deployment scenario requirements with regard to confidentiality and maximizes availability.

Parts of the contributions of this thesis have been previously published in:

- Jens Köhler, Konrad Jünemann, and Hannes Hartenstein. *Confidential database-as-a-service approaches: taxonomy and survey.* Journal of Cloud Computing, vol. 4, pages 1-14, 2015.
- Jens Köhler and Konrad Jünemann. *Securus: From confidentiality and access requirements to data outsourcing solutions.* In Privacy and Identity Management for Emerging Services and Technologies, IFIP Advances in Information and Communication Technology, vol. 421, pages 139–149. Springer Berlin Heidelberg, 2014.
- Jens Köhler, Konrad Jünemann, and Hannes Hartenstein. *Securus: Composition of confidentiality preserving indexing approaches for secure database-as-a-service.* PIK-Praxis der Informationsverarbeitung und Kommunikation, vol. 37, pages 149–155, 2014.
- Konrad Jünemann, Jens Köhler, and Hannes Hartenstein. *Data outsourcing simplified: Generating data connectors from confidentiality and access policies.* In Proceedings of the IEEE/ACM Workshop on Data-intensive Process Management in Large-Scale Sensor Systems (DPMSS-CCGrid), 2012.
- Jens Köhler and Hannes Hartenstein. *Index optimization for ℓ-diversified database-as-a-service.* In Proceedings of the 9th International ESORICS Workshop on Data Privacy Management (DPM-ESORICS), 2014.
- Jens Köhler and Hannes Hartenstein. *Occasio: an operable concept for confidential and secure identity outsourcing.* In Proceedings of the IFIP/IEEE International Symposium on Integrated Network Management (IM), 2013.
- Jens Köhler, Sebastian Labitzke, Michael Simon, Martin Nussbaumer, and Hannes Hartenstein. *Facius: An easy-to-deploy SAML-based approach to federate non web-based services.* In Proceedings of the 11th IEEE International Conference on Trust, Security and Privacy in Computing and Communications (TrustCom), 2012.
- Jens Köhler, Michael Simon, Martin Nussbaumer, and Hannes Hartenstein. *Federating HPC access via SAML: Towards a plug-and-play solution.* In Proceedings of the International Supercomputing Conference, Lecture Notes in Computer Science, vol. 7905, pages 462–473. Springer Berlin Heidelberg, 2013.
- Jens Köhler, Sebastian Labitzke, Michael Simon, Tobias Dussa, Martin Nussbaumer, and Hannes Hartenstein. *bwIDM – Federated access to IT-based services at the universities of the state of Baden-Württemberg.* PIK-Praxis der Informationsverarbeitung und Kommunikation, vol. 37, pages 15–21, 2014.
- Jens Köhler, Jens Mittag, and Hannes Hartenstein. *User-centric management of distributed credential repositories: Balancing availability and vulnerability.* In Proceedings of the 18th ACM Symposium on Access Control Models and Technologies (SACMAT), pages 237–248, New York, NY, USA, 2013. ACM.

1.5 Structure of the Thesis

The thesis is structured as follows. We introduce the concept of tunability and the methodology on how to build tunable approaches in Chapter 2. We apply this methodology to the research challenge of secure database outsourcing and introduce the resulting tunable approaches in Chapter 3. In Chapter 4, we address the research question we stated for secure identity outsourcing. The research question we stated for secure credential outsourcing is addressed in Chapter 5. Finally, we provide a conclusion and an outlook in Chapter 6.

2
Methodology

In this chapter we define precisely the concept of tunability and propose a methodology on how to build tunable approaches. Furthermore, we place approaches with tunable security properties in context with other research fields and established standards for software and system quality as well as for security management and secure system development.

In this thesis we make use of the terminology that is introduced by the ISO 25000 standard [ISO14a] for modeling software and system quality. We provide an overview of the ISO 25000 quality models in Section 2.1. As the ISO 25000 standard addresses quality models rather generic without focusing on security, we provide an overview of the ISO 270xx [ISO14b] information security management process and the KASTEL method [Kas14], a unifying process to develop secure systems in Section 2.2. We use this as a basis to place our contributions. We motivate and define the concept of tunable approaches in Section 2.3 and illustrate their potential through a case study in Section 2.4. We provide a generic methodology on how to develop tunable approaches in Section 2.5 and show how findings of other research fields including operations research and policy-based management can be leveraged within this methodology. Furthermore, we provide an overview of the domain specific trade-offs that we address in the remaining chapters of this thesis by applying the methodology to build approaches which allow to tune the trade-offs. We summarize the methodology chapter in Section 2.6.

2.1 Software Quality

The ISO/IEC 25000 standard [ISO14a] defines **software quality** as the "capability of software product to satisfy stated and implied needs when used under specified conditions". In particular, this definition implies that the quality of specific software is *not* to be understood as a fixed value but has to be evaluated for each context in which the software is used.

System/Software Product Quality Model

Functional Suitability	Functional Completeness
	Functional Correctness
	Functional Appropriateness
Performance Efficiency	Time Behavior
	Resource Utilization
	Capacity
Compatibility	Co-existence
	Interoperability
Usability	Appropriateness Recognisability
	Learnability
	Operability
	User Error Protection
	User Interface Aesthetics
	Accessibility
Reliability	Maturity
	Availability
	Fault Tolerance
	Recoverability
Security	Confidentiality
	Integrity
	Non-repudiation
	Accountability
	Authenticity
Maintainability	Modularity
	Reusability
	Analysability
	Modifiability
	Testability
Portability	Adaptability
	Installability
	Replaceability

Quality In Use Model

Effectiveness	Effectiveness
Efficiency	Efficiency
Satisfaction	Usefulness
	Trust
	Pleasure
	Comfort
Freedom from risk	Economic Risk Mitigation
	Health and Safety Risk Mitigation
	Environmental Risk Mitigation
Context coverage	Context Completeness
	Flexibility

Figure 2.1: Quality models according to the ISO/EIC 25010 standard [ISO11b].

Definition - Context of Use [ISO14a]: A context of use is defined by ISO/IEC 25000 as "users, tasks, equipment (hardware, software and materials), and the physical and social environments in which a product is used" [ISO14a].

To enable a structured evaluation of software quality, the ISO/IEC 25010 standard defines **software quality models** as a "defined set of characteristics, and of relationships between them, which provides a framework for specifying quality requirements and evaluating quality" [ISO11b].

Definition - Quality Characteristic [ISO11b]: A quality characteristic is defined as a "category of software quality attributes that bears on software quality" [ISO11b].

Definition - Quality Property [ISO11b]: A quality property is defined as a "measurable component of quality" [ISO11b].

The ISO/IEC 25010 standard distinguishes between a *product quality* model and a *quality in use* model. While the characteristics of the product quality model can be applied to the software product as such and the target computer system on which the software is executed, the quality in use model can be applied to assess the "degree to which a product or system can be used by specific users to meet their needs to achieve specific objectives with effectiveness, efficiency, freedom from risk and satisfaction in specific contexts of use" [ISO14a].

The defined characteristics of the product quality and the quality in use model are shown in Figure 2.1. It can be distinguished between characteristics such as "Security" and subcharacteristics such as "Confidentiality" and "Integrity" that allow to assess the characteristic in a more fine grained manner. It's important to note that not all characteristics are relevant for every stakeholder and in every contexts of use [ISO11b]. The specified models are applicable to software products and computer systems in general. In fact many of the specified characteristics are also relevant to systems other than software/computer systems [ISO11b].

To assess the quality properties of a system with regard to the characteristics of the proposed quality models, measurement methods and models are proposed in the ISO/IEC 2502n standards [ISO07b] and ISO/IEC 15939 [ISO07a]. Furthermore, methodologies to specify quality requirements are introduced in the ISO/IEC 2503n standard series [ISO07c]. Based on specified requirements that depend on the context of use, the software quality of an existing software product can be evaluated by using the methodologies presented in ISO/IEC 25040 [ISO11c]. While the ISO 250xx standards provide a framework to assess software quality quantitatively and to check whether quality requirements are met, quantifying certain properties is hard and not practical in many cases. For instance, developing metrics to measure confidentiality or anonymity is still an open field of research [FWCY10].

In this thesis, we do *not* focus on quantifying systems' quality properties but investigate interdependencies between quality characteristics and show how trade-offs between different quality characteristics can be achieved.

2.2 System Security

Security is one characteristic of software and system quality. The ISO 25000 standard defines security to be the "degree to which a product or system protects information and data so that persons or other products or systems have the degree of data access appropriate to their types and levels of authorization" [ISO14a]. For a system to be considered secure, it suffices to guarantee that the risk of unauthorized access is acceptable [San03, ISO14b]. Furthermore, not every risk needs to be addressed. For instance, it makes no sense to address the risk of an unauthorized access if addressing the risk is more expensive than the occurrence of the unauthorized access. Managing IT security risks, i.e., deciding on how to cope with risks is important and constitutes a field of research on its own.

In the following, we give an introduction to IT security risk management based on the ISO 270xx standard series [ISO14b, ISO11d]. To provide an understanding of the development of secure systems, we provide an overview of the KASTEL method [Kas14], a unifying framework to guide the development of secure systems. We show how it relates to the generic IT security management workflow and how our contributions relate to it.

IT Security Management

A high-level overview of managing IT security according to the ISO 270xx standard series [ISO14b, ISO11d] is shown in Figure 2.2.

Step 1 - Identification of security objectives: In a first step, the security objectives have to be identified. Security objectives can be the protection of high value information assets, the business needs for information processing, storage, and communication as well as the fulfillment of legal, regulatory, and contractual requirements.

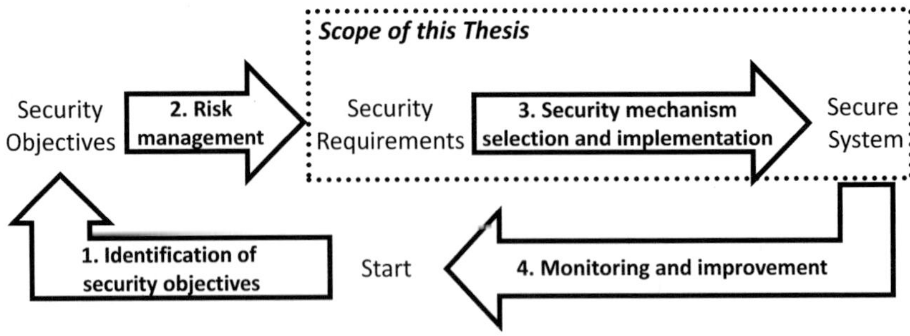

Figure 2.2: IT Security Management Process according to ISO 270xx standard series [ISO14b, ISO11d].

Step 2 - Risk management: A risk manager has to identify, analyze and assess risks with regard to achieving the specified security objectives. Based on the assessment, risks have to be reduced to an acceptable level [ISO11d]. Risk treatments include risk avoidance, e.g., not outsourcing data at all, risk retention, e.g., simply accepting the risk of data disclosure, risk sharing, e.g., sharing the risk with other entities such as insurance companies, and risk mitigation, e.g., implementing measures to reduce the risk. In this thesis, we focus on risk mitigation by requiring systems to enforce specific security properties. For instance, a requirement could be to enforce the confidentiality of a specific piece of data to achieve a higher level security objective. The data confidentiality requirement can be enforced by applying security mechanisms such as encryption to achieve specific security properties.

> *Definition - Security Mechanism:* A security mechanism enforces specific security properties on a technical level and can be used to satisfy security requirements.

By applying security mechanisms, the cost of undermining security can be increased from an attacker's point of view. This reduces risk as attackers are less likely to perform an attack. The higher the cost from an attacker's point of view, the smaller the occurrence probability, the smaller the risk. Note that security requirements do not necessarily dictate which security mechanisms have to be applied. For instance, the confidentiality of the data can also be enforced by storing it on a trusted system instead of encrypting it.

Thus, the risk manager can mitigate the risks to a level that is considered appropriate by specifying which security requirements a system has to satisfy to be applied in a given context of use [MLY05]. While security objectives can be understood as business goals that can be ambiguous, security requirements constitute unambiguous requirements which can act as input for system development processes.

Step 3 - Security mechanism selection and implementation: Security mechanisms are selected that satisfy the specified security requirements. The selected security mechanisms are implemented to build a secure system that satisfies the specified requirements. In this context, a system is considered "secure" if it correctly enforces the security requirements that were specified by the risk management.

Step 4 - Monitoring and improvement: As it is common practice in IT security management processes [ISO14b], it has to be continuously checked whether the system is still suited for the current context of use, i.e., it has to be monitored whether the deployment scenario changes. For instance, in case more powerful attackers have to be addressed, the risk assessment has to be performed again and in case security requirements changed the secure system has to be re-developed.

The scope of this thesis is the development of technical solutions that satisfy requirements which reflect the decisions made by risk management. This is also part of secure system development workflows. To provide a more in-depth view on secure system development we provide an overview of the KASTEL method [Kas14] and show how our contributions relate to it.

Secure System Development - The KASTEL Method

The KASTEL method [Kas14] constitutes a unifying framework to guide the development of secure systems. An overview of the KASTEL method is shown in Figure 2.3. In particular, the method highlights the factors of influence on system security and points out that developing secure systems is not only about the implementation of security mechanisms that enforce specific security properties. Instead, security can break at each part of the software development process, including the system specification, system design and system implementation.

The KASTEL method proposes the following process to develop secure systems:

Step 1 - Requirements Elicitation: In the first step, a stakeholder defines security requirements on the system on an abstract, imprecise level. This can be seen as analogon to step 1 of the general IT security management workflow.

Step 2 - Knowledge Engineering: From the initial abstract and possibly imprecise requirements specification, an accurate and unambiguous system specification is derived that contains accurate security requirements that have to be met to satisfy the ambiguous and possibly imprecise security requirements specified by the stakeholder. In this step, it is decided which specific security property requirements a system has to satisfy to be considered "secure". To determine which specific security requirements have to be satisfied, a variety of factors have to be taken into account such as existing trust relationships, law, and domain specific knowledge on which vulnerabilities can be exploited.

> *Example:* If a stakeholder requires data confidentiality and the party that stores the data is trusted to maintain data confidentiality, there is no need to additionally enforce confidentiality by the technical system.

The Knowledge Engineering step is comparable to the risk management step (step 2) in the general IT security management workflow in the sense that the requirements that have to be satisfied for a system to be secure in a given context of use are precisely defined.

Step 3 - System Design: Based on the system specification, a system design is created. The system design includes choosing security mechanisms that enforce the security properties as required in the system specification.

Step 4 - System Implementation: Finally, the set of security mechanisms contained in the system design is implemented to build the final system. System design and implementation can be considered as part of the development step (step 3) of the general IT security management workflow.

For the final system to be "secure", i.e., to satisfy the stakeholders requirements, steps 2-4 have to be performed correctly, i.e., a system specification that *correctly* enforces the stakeholder requirements specification has to be determined based on knowledge engineering, a system design that *satisfies* the system specification has to be determined and it has to be *correctly* implemented. Errors in any of these three steps

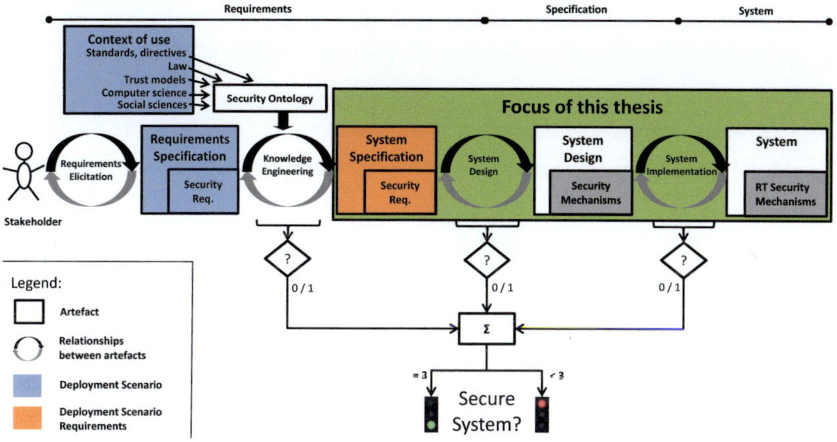

Figure 2.3: Development of secure systems according to the KASTEL method (adapted from [Kas14]).

result in an insecure system. For instance, if requirements are incorrectly specified in step 2, the final system has to be considered insecure as it is only guaranteed to enforce the incorrect requirements. If the system is designed correctly, but the implementation does not correctly reflect the system design, the system cannot be guaranteed to be secure. If the requirements are correctly specified and the implementation of the system design is correct but the system design does not meet the system specification, the security of the final system cannot be guaranteed.

2.3 Deployable & Tunable Approaches

At first glance, the KASTEL method appears to be a finite process. However, it has to be considered as a part of the IT security management process to take evolving security ontologies and stakeholder requirements into account. In the knowledge engineering step (2) of the KASTEL method, both the requirements specification and the security ontology can change over time. This may even be the case during the design and implementation phase of the system. In such cases, the system specification has to be adapted and it is not guaranteed that the old system implementation still satisfies the requirements of the new system specification and is deployable in the deployment scenario.

> *Definition - Deployment scenario:* We define a deployment scenario of a system as a combination of the stakeholder's requirements specification and a given context of use.

> *Definition - Deployment scenario requirements:* We define deployment scenario requirements to be an instance of a system specification. Deployment scenario requirements comprise requirements with regard to both security and other quality characteristics.

Definition - Deployable: A system is *deployable* in a specific deployment scenario if it satisfies the deployment scenario requirements.

Whether a system can be considered deployable in a given deployment scenario depends on 1) whether it satisfies the scenario's security requirements and 2) whether it satisfies the scenario's requirements with regard to other quality characteristics such as performance efficiency. The quality properties of a system determine the system's area of application.

Definition - Area of application: A system's area of application is determined by the set of deployment scenarios in which the system is deployable.

Maximizing a system's area of application is beneficial. Systems with a wide area of application address a bigger market from a business perspective. From an open source perspective, systems with a wide area of application have a larger user base which increases participation. Maximizing a system's area of application is also beneficial due to the fact that deployment scenario requirements may change over time due to evolving stakeholder requirements or an evolving context of use (e.g., trust relationships that change over time). Approaches that satisfy the old and the new deployment scenario requirements and are also deployable in the new deployment scenario are desirable. This is especially relevant for software development, where initial requirements and contexts of use may change even before the approach is deployed, i.e., during the design and implementation phase [Che03].

To maximize the area of application, system developers aim to satisfy as many deployment scenario requirements as possible by maximizing the system's quality properties. However, maximizing two quality properties is not possible if trade-offs between them exist.

Definition - Quality characteristics trade-offs: A trade-off between quality characteristics exist in a given problem setting if multiple characteristics depend on each other and improving one characteristic impairs another quality characteristic.

If quality characteristics trade-offs have to be made in a given problem setting, deployment scenarios requirements can be conflicting, i.e., it is not possible to satisfy them at once. Thus, maximizing each quality property to satisfy all possible scenario requirements is impossible. This is illustrated in the following example.

Example: Consider the trade-off between efficiency and confidentiality in the problem setting of outsourcing data to a cloud provider. The trade-off situation is illustrated in Figure 2.4 based on possible scenario requirements and approaches with different quality properties.

In deployment scenario X the requirements are that data confidentiality has to be enforced by technical means as the cloud provider is not considered trustworthy. In deployment scenario Y, the ability to efficiently search in the outsourced data is required.

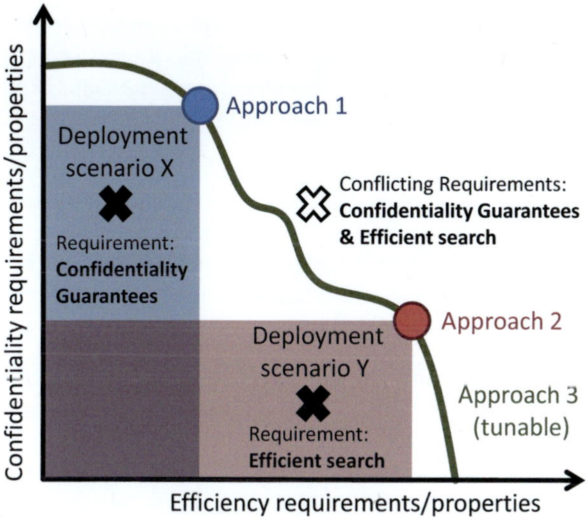

Figure 2.4: Example: The trade-off situation between confidentiality and efficiency in the database outsourcing problem setting.

> Approach 2 outsources plaintext data enables efficient searches on the data, but does give the provided confidentiality guarantees as the cloud provider can read the data. To satisfy the confidentiality requirements, approach 1 encrypts the data before it is outsourced. However, this makes searches inefficient as the cloud provider cannot efficiently search encrypted data that is indistinguishable.
>
> Satisfying both requirements, guaranteed confidentiality and efficient searches, is not possible based on known security mechanisms. Thus, the requirements of efficient search and confidentiality guarantees are conflicting and no approach can satisfy both requirements at once.
>
> However, we show in this thesis that the trade-off can be adjusted in a more fine-grained manner by applying security mechanisms that allow *specific* search operations to be executed efficiently and in exchange leak information about *parts* of the outsourced data. Thus, by relaxing the efficiency and confidentiality requirements a bit, the requirement conflict can potentially be resolved.

The example shows that quality characteristics trade-offs that are inherent to the problem setting force systems to accept a lower level of one quality characteristic in order to achieve a better level for another quality characteristic. Hard-coding these trade-off choices in the system's implementation limits the system's area of application and risks that the approach can no longer be applied if deployment scenario requirements change over time.

Example: The example shown in Figure 2.4 shows that hard-coding trade-off decisions in the systems implementation limits the area of application. Approach 1 satisfies the deployment scenario requirements in the blue area and approach 2 satisfies those in the red area. Approach 1 satisfies the requirements of deployment scenario X, while approach 2 does not.

Ideally, to maximize a system's area of application it should be *tunable* in the sense that characteristics can be traded for each other without having to re-design and re-implement the system. Thus, a tunable system can be adapted to scenario requirements and covers more deployment scenario requirements than an approach that hard-coded a specific trade-off choice does.

Definition - Tunable approaches: We consider an approach tunable w.r.t. quality characteristics $\mathcal{C} = \{C_1,\ldots,C_n\}$ if the approach can be parameterized to satisfy requirements with regard to the characteristics $\mathcal{R} \subseteq \mathcal{C}$ and maximize the quality levels for the characteristics of $\mathcal{M} \subseteq \mathcal{C}$. In particular, the approach can achieve better levels of quality for the characteristics $\mathcal{M}$ if the requirements on the characteristics $\mathcal{R}$ are less tight.

Example: In Figure 2.4, approach 3 is tunable and can tune its confidentiality and efficiency properties to those of approach 1 and 2. Thus, it is deployable in all deployment scenarios covered by approach 1 and 2, including deployment scenario X and Y. Furthermore, approach 3's area of application exceeds that of approach 1 and 2 as it can additionally satisfy the requirements of deployment scenarios that are beneath the green graph which are not covered by the blue and the red box.

The concept of tunability is not limited to *quantitative* requirements that can be measured by a continuous metric such as time behavior in milliseconds, but can also be applied for *qualitative* requirements such as "users should not be obligated to download software in order to use the approach". Furthermore, it is not necessary that requirements regarding characteristic are comparable in the sense that they can be ordered. For instance, an approach can also be tunable in the sense that it matches (non-orderable) confidentiality requirements like "Attribute A and B must not be viewable together" with the requirement "query X should be executable in at most 300 ms".

In this thesis we identify trade-offs between security sub-characteristics and selected other quality characteristics, show how they can be tuned, and propose approaches that allow to automatically tune the trade-off to meet deployment scenario requirements. In other words, we explore how tunable approaches can be developed that have a large area of application by being tunable to deployment scenario requirements. The tunable approaches we propose for selected data outsourcing problem settings automate the process shown in Figure 2.3 from step 2 on. Thus, our approaches can be applied instead of designing and implementing a new system that is tailored to satisfy the requirements of a given deployment scenario. To apply them for a given deployment scenario, only the deployment scenario requirements have to be determined based on the stakeholders requirements and the context of use.

2.4 Case Study: The bwIDM Identity Federation

To illustrate the relevance of tunable systems in practice, we show how Facius [KLS+12, KSNH13], a tunable approach we developed to integrate services with identity federations, is applied in practice within the bwIDM identity federation.

In the state of Baden-Württemberg, state-funded academic IT services such as high performance computing (HPC) clusters, large file storage systems and library services are consolidated at single universities (see Figure 2.5) [HWC13]. As the services are funded by the state, they should be accessible by *all* academic users that are affiliated with a university of Baden-Württemberg. For instance, a user A that is affiliated with university X should be able to utilize the HPC cluster of university Y.

If traditional intra-organizational identity management solutions were used, a user that wants to access an IT service of a foreign university would have to create a local account at this university. This account is independent from the user's account at her home organization, i.e., the university the user is affiliated with. Thus, the user has to manage an additional set of credentials for the newly created local account and has to maintain and update identity information such as residential or contact e-mail addresses once they change. In the worst case, a user that wants to access multiple services that are hosted by n universities has to manage and maintain n accounts.

In the context of the bwIDM project [KLS+14], an identity federation was established that enables users to access the state-funded services based on the users' home-organizational accounts without being required to establish local accounts at each service. The bwIDM federation is based on the Security Assertion Markup Language (SAML) Standard [HCH+05] to make use of the existing SAML identity providers. This kind of IdPs are already operated within the DFN-AAI federation[1] at a majority of Baden-Württemberg's universities to federate access to web-based services such as the Elsevier[2] or Springer Link[3] library services. Integrating the non web-based services of the bwIDM federation with these IdPs is desirable to avoid having to build application-specific identity federations from scratch and to leverage existing trust relationships. The integration of existing non web-based services into the bwIDM federation was performed based on the tunable Facius approach [KLS+12] that we developed within the bwIDM project.

Integrating non web-based services with existing web-based federations constitutes a challenge as trade-offs exists between security characteristics and other characteristics such as service availability and usability. At the user's end, traditional non web-based service clients (e.g., SSH clients) send the user's password directly to the service provider (e.g., the SSH server) instead of the identity provider. At the service provider, many established non web-based services are designed for traditional inner-organizational identity management solutions like *Lightweight Directory Access Protocol* (LDAP) servers and do not support federated authentication and authorization. Adapting existing proprietary services to support federations is not feasible in many scenarios. While we designed the Facius approach to be made transparent

[1] https://www.aai.dfn.de/ [last visited on March 2015]

[2] http://www.elsevier.de/ [last visited on March 2015]

[3] http://link.springer.com/ [last visited on March 2015]

to the service by providing a common LDAP interface [KSNH13] to the service, it is inherently impossible to make service clients send the user password to the IdP instead of the SP without modifying them. Thus, there exists a trade-off between usability, i.e., allowing users to use their accustomed service clients, and security, i.e., not enabling the SP to view user passwords.

In the following, we show on a conceptual level how different deployment scenario requirements can exist based on the existing trust relationships within a federation and how security mechanisms to satisfy those requirements influence usability and availability characteristics. For a more detailed view on trust relationships in federations and deployment scenario requirements we refer to Section 4.2. For a more detailed and more technical description of the Facius approach see [KLS+12].

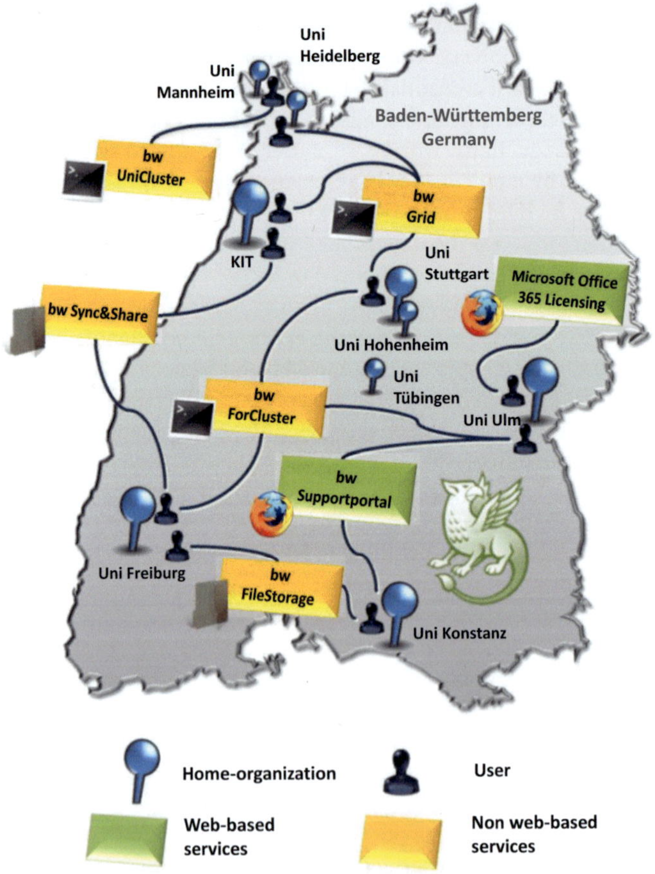

Figure 2.5: Overview of the bwIDM federation.

Deployment scenario 1: Service providers are not trusted by the IdPs and the users to handle user passwords, i.e., the risk of an attacker compromising an SP and eavesdropping passwords is considered too high. The Facius approach provides two options to avoid the risk of password interception at the SPs: 1) use an enhanced service client that sends to password directly to the IdP and relays the IdPs signed confirmation that the user correctly authenticated to the SP and 2) authenticate the user locally based on either an SP-specific password that was established during the registration for the service or a public-key that was deployed during the registration to enable a challenge response authentication. Both options can decrease usability. Users can be confused if it is not possible to use their accustomed home-organizational password to log on. Likewise, downloading a modified service client only to access one specific service can be considered cumbersome if non-modified clients such as standard SSH clients are already installed. Notice that in some deployment scenario such as the Dropbox-like file synchronization solution bwSync&Share a new service client has to be installed by the user anyway. Thus, in this case, the modified service client option does not decrease usability. □

Deployment scenario 2: Service providers are trusted by the IdPs and users to handle user passwords. The risk of the compromised SPs that eavesdrop passwords is considered acceptable. Based on this trust relationship, sending the home-organizational password to the service is feasible. The service can then validate the password against the LDAP interface of Facius like it would with a local identity management. The Facius component behind the LDAP interface then sends the password to the IdP to validate it and notifies the service of the outcome. Thus, Facius can leverage the trust relationship in this deployment scenario to increase usability as users are able to login to services with their familiar credentials and unmodified service clients. □

Deployment scenario 3: Service providers are not trusted by the IdPs and users to handle user passwords. Furthermore, authorization decisions have to be strictly correct and must not be based on outdated authorization tokens, i.e., the freshness of authorization tokens is paramount when making authorization decisions. For instance, it must not be possible for a user that recently lost the group membership "admin" to gain admin rights at an SP. Facius allows to enforce this security requirement by strictly forcing the retrieval of authorization tokens from the IdP each time an authorization decision has to be made. Even in case the user authenticated against the SP using an SP-specific password, the IdP of the user is queried for up-to-date user attributes. However, retrieving authorization tokens from the IdP each time an authorization decision has to be made harms the availability of the service. If the IdP is not available, none of its users can access services as the services are not able to retrieve up-to-date authorization tokens. □

The use of Facius in the deployment scenarios shows that by tuning security characteristics (e.g., the integrity/freshness guarantees of authorization tokens and the protection against SPs that eavesdrop passwords) higher levels of usability and availability can be achieved. Compared to static approaches that cannot be configured to tune security characteristics, the Facius approach can satisfy the requirements of a bigger variety of deployment scenarios.

In the exemplary case of the bwIDM federation, this is an important feature, as some home organizations only trust selected SPs to handle user passwords. Furthermore, the SP requirements with regard to the freshness of authorization tokens vary. For instance, when accessing a HPC service, for each login it has to be guaranteed that the user still possesses the according rights. This is not a pressing requirement for other services for which the guarantee suffices that the user possessed the according rights during the past x hours. The more SPs and IdPs participate in a federation, the higher the heterogeneity with regard to security requirements will become. This underlines the importance of approaches with tunable security properties like Facius, as otherwise the SP/IdP with the highest security requirements would dictate the necessity to apply security mechanisms and all other SPs/IdPs would have to accept the negative effects of these security mechanisms on other quality characteristics.

2.5 Building Tunable Approaches

To address the fundamental research question we propose a generic methodology on how tunable approaches can be developed. We apply this generic methodology to build tunable approaches that address various domain specific problems in the remainder of this thesis. To build a tunable approach, the following steps have to be performed.

Step 1 - Identification of trade-offs: Designing tunable approaches only makes sense if inherent quality characteristic trade-offs have to be made within a problem setting. Otherwise, an approach that maximizes all quality characteristics can be designed which can be deployed in all deployment scenarios. In a first step, these quality characteristic trade-offs have to be identified. We show in Section 2.5.1 which trade-offs are addressed by the tunable systems that are proposed in this thesis.

Step 2 - Identification and assessment of security mechanisms: To satisfy the deployment scenario security requirements, security properties have to be enforced by security mechanisms. These security mechanisms have to be identified and assessed with regard to the provided level of protection and their impact on other quality characteristics. Based on the assessment it can be determined how the identified trade-offs can be made and tuned.

Step 3 - Definition of a deployment scenario requirements language: The deployment scenario requirements have to be specifiable by domain experts of the approach's application domain. These domain experts do not necessarily have expert knowledge on the security mechanisms that can be used to enforce their deployment scenario requirements. Thus, a language that abstracts as much as possible from the security mechanisms is required to express deployment scenario requirements. We show how this relates to the concept of policy-based management in Section 2.5.2.

Step 4 - Development of a transformation method to match security mechanisms with deployment scenario requirements: Based on the specified deployment scenario requirements, a tunable approach should be capable of automatically determining and implementing a suitable set of security mechanisms that are suitable to enforce the security requirements and optimize other quality characteristics. As we show in this thesis, in some cases the resulting optimization problem can be trivially solved by providing a mapping of deployment scenario requirements on security mechanisms. However, in some problem settings the optimization problem is complex. We provide fundamentals on solving optimization problems in Section 2.5.3.

In the remaining chapters of this thesis, we show how the methodology can be applied to develop tunable approaches that address domain specific problems in the domains of database outsourcing, identity outsourcing and credential outsourcing. We show which trade-off situations are addressed by our contributed tunable approaches in Section 2.5.1. In particular, the tunable approaches that we provide in this thesis show how methods of policy-based management (Section 2.5.2) and operations research (see Section 2.5.3) can be leveraged within the proposed methodology.

2.5.1 Examined Trade-Offs

Like the ISO/EIC 25010 standard, we do not claim to address every quality characteristic in this thesis. In this section we provide an overview of the quality characteristics that are relevant to understand the contents of this thesis and show which trade-offs our proposed tunable approaches address for each problem setting. The problem settings we address in this thesis are database outsourcing, identity outsourcing and credential outsourcing. An overview of the trade-offs between security and other quality characteristics that are addressed in this thesis is shown in Figure 2.6.

Security is defined as the "degree to which a product or system protects information and data so that persons or other products or systems have the degree of data access appropriate to their types and levels of authorization" [ISO11b]. The security characteristics that we investigate in this thesis include the following:

- **Confidentiality**: "degree to which a product or system ensures that data are accessible only to those authorized to have access" [ISO11b].
- **Anonymity**: degree to which a product or system ensures that data cannot be mapped to an individual[4].
- **Integrity**: "degree to which a system, product or component prevents unauthorized access to, or modification of, computer programs or data" [ISO11b].

We investigate trade-offs between the listed security characteristics and the characteristics of *performance efficiency*, *reliability*, *usability* and *compatibility*. In the following we provide the ISO 25010 definition for each characteristic and give an overview of the trade-offs we identified and addressed in this thesis. For a more detailed analysis of each identified trade-off we refer to the section in which the tunable approach that addresses the trade-off is introduced.

[4] The ISO 25010 does not specify an anonymity characteristic. However, we feel that it is an important security sub-characteristic that is not covered by the other sub-characteristics that are defined in the ISO 25010 standard. This point of view is also shared by other established researchers in the community [Eck06].

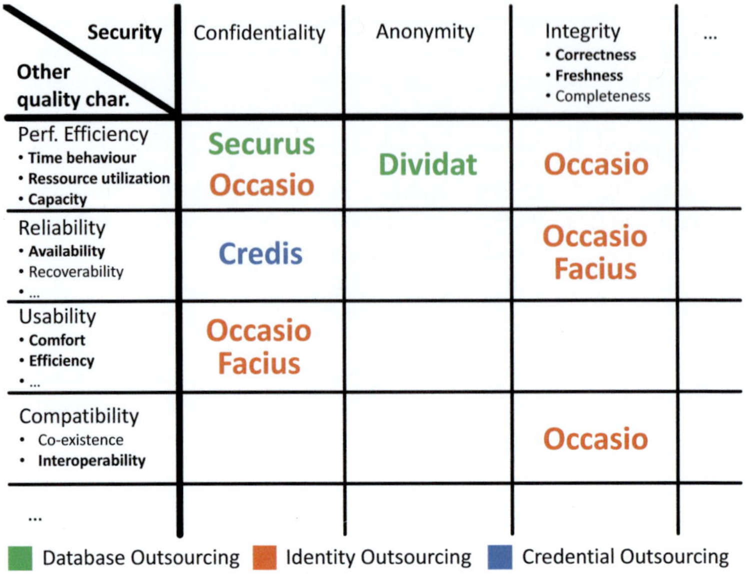

Figure 2.6: Examined quality characteristics in this thesis.

Trade-offs between Performance Efficiency and Security

Performance efficiency is defined as "performance relative to the amount of resources used under stated conditions" [ISO11b]. Enforcing security properties by mechanisms such as encryption can negatively affect the efficiency of an approach. Subcharacteristics of performance efficiency according to ISO 25010 include:

- **Time behavior**: "degree to which the response and processing times and throughput rates of a product or system, when performing its functions, meet requirements" [ISO11b].
- **Resource utilization**: "degree to which the amounts and types of resources used by a product or system, when performing its functions, meet requirements" [ISO11b].
- **Capacity**: "degree to which the maximum limits of a product or system parameter meet requirements" [ISO11b].

When outsourcing databases to potentially curious storage providers, the confidentiality of the database records can be enforced by encrypting the records so that they are indistinguishable from one another. However, as the records cannot be distinguished anymore, the database can no longer be queried for records that match specific conditions. As the whole database has to be decrypted before a query can be evaluated, efficiency overheads for query executions are incurred. To enforce anonymity, anonymized data can be outsourced. However, executing database queries on anonymized data that return the same results as they would if they were executed on the original data can incur efficiency overheads as we show in this thesis. We propose the *Securus* and *Dividat* approaches that can be tuned to satisfy confidentiality and anonymity requirements and optimize efficiency in terms of query latency, transmission and storage overhead at the same time.

In the case of identity outsourcing, some deployment scenarios require the enforcement of freshness, i.e., the guarantee that a specific identity is not outdated and no newer version of it exists already. Furthermore, some deployment scenarios require the enforcement of the confidentiality of identity information against potentially malicious hosters that store the identities. We show that enforcing confidentiality and freshness incurs efficiency overheads and propose the *Occasio* framework which can be tuned to enforce both freshness and confidentiality at the expense of increased resource utilization.

Trade-offs between Reliability and Security

Reliability is defined as the "degree to which a system, product or component performs specified functions under specified conditions for a specified period of time" [ISO11b]. Enforcing security properties can negatively affect the reliability of an approach. Sub-characteristics of reliability according to ISO 25010 include:

- **Availability**: "degree to which a system, product or component is operational and accessible when required for use" [ISO11b].
- Reliability includes further sub-characteristics such as maturity, fault tolerance, and recoverability which are not addressed in this thesis.

The confidentiality of outsourced credentials can be enforced by splitting them up on multiple non-colluding parties. All of those parties have to be compromised to access the credentials in an unauthorized way. However, also all of the parties have to be available to retrieve the credentials in an authorized way. Thus, there exists a trade-off between confidentiality and availability. We propose the *Credis* framework to outsource credentials that maximizes the availability of the credentials by tuning the provided level of confidentiality to the deployment scenario requirements.

In federated identity management, freshness and availability constitute conflicting objectives. To be absolutely certain that the identity information on which an authorization decision is based is up to date, the identity information has to be attested by the home organization that is the authoritative source of it. However, if the home organization has to attest the freshness of a user identity each time the user logs on, every relying service provider is not available if the home organization is unavailable. In this thesis, we propose the *Occasio* approach which allows to maximize availability by tuning the provided freshness guarantees to the level that is required in a given deployment scenario. Furthermore, the *Facius* approach presented in Section 2.4 can be tuned to either query the users identity from the home organization each time the user logs on if freshness is required or to rely on identity data that was cached at a previous login.

Trade-offs between Usability and Security

Usability is defined as the "degree to which a product or system can be used by specified users to achieve specified goals with effectiveness, efficiency and satisfaction in a specified context of use" [ISO11b]. Enforcing security properties can lead to a decreased level of usability. Sub-characteristics of usability according to ISO 25010 include:

- **Efficiency**: "resources expended in relation to the accuracy and completeness with which users achieve goals" [ISO11b]
- **Satisfaction**: "degree to which user needs are satisfied when a product or system is used in a specified context of use " [ISO11b]
- Usability included further sub-characteristics such as learnability, operability and user interface aesthetics which are not addressed in this thesis.

In the problem setting of federated identity management, enforcing the confidentiality of user credentials against SPs that are not trusted to view user credentials can negatively affect usability (see Section 2.4). To enforce the confidentiality of credentials, different authentication mechanisms such as challenge response protocols or SP-specific passwords can be used. Thus, users have to synchronize cryptographic keys on their end devices or they have to memorize a password for each SP. Furthermore, clients that support authentication against IdPs instead of SPs can be used. If the users already have clients installed that do not support this form of authentication they have to install a modified version which constitutes usability overhead. Depending on the deployment scenario, both options can constitute a significant overhead for users [HVO12] and considerably decrease the users' efficiency and satisfaction. The *Facius* approach supports multiple authentication methods and can be tuned to meet deployment scenario requirements with regard to credential confidentiality to optimize usability by only requiring the user to use modified service clients or alternative authentication methods if absolutely necessary.

With *Occasio* we propose an approach that can be used by home organizations to securely outsource their IdP to highly available, untrusted providers. Thus, home organizations that do not possess a highly available infrastructure can still offer their users a highly available access to relying services. However, this problem setting induces a trade-off between usability and credential confidentiality. The SP has to perform the authentication as the home organization cannot be assumed to be available and the party that stores the identities is not trusted to view user credentials. Similarly to the Facius problem setting (see Section 2.4) usability is negatively affected if the SP is not trusted to view user credentials as home-organizational credentials can no longer be used without revealing them to the SP. The *Occasio* approach supports multiple authentication methods and can be tuned to meet deployment scenario requirements with regard credential confidentiality to optimize usability by only requiring the user to use modified service clients or alternative authentication methods if absolutely necessary.

Trade-offs between Compatibility and Security

Compatibility is defined as the "degree to which a product, system or component can exchange information with other products, systems or components, and/or perform its required functions, while sharing the same hardware or software environment" [ISO11b]. Enforcing security properties can negatively affect the compatibility of an approach. Sub-characteristics of compatibility according to ISO 25010 include:

- **Co-existence**: "degree to which a product can perform its required functions efficiently while sharing a common environment and resources with other products, without detrimental impact on any other product" [ISO11b]
- **Interoperability**: "degree to which two or more systems, products or components can exchange information and use the information that has been exchanged" [ISO11b]

We show that based on techniques that are currently supported by prevalent federated identity management standards, guaranteeing freshness in the problem setting of Occasio incurs an unbearable efficiency overhead in most deployment scenarios. Alternative techniques induce substantially less overhead and make the enforcement of freshness feasible. However, as they are not supported by commonly deployed SP implementations, modified SP implementations are required. Thus, applying efficient techniques to enforce freshness is not interoperable with common, unmodified SPs. Depending on the deployment scenario requirements, *Occasio* can be tuned to make use of the alternative techniques if freshness is required or to support unmodified SPs if freshness is not required.

2.5.2 Fundamentals: Policy-Based Management

The deployment scenario requirements have to be specifiable by the risk manager and domain experts of the approach's application domain. The domain experts and the risk manager do not necessarily have expert knowledge on the security mechanisms that can be used to enforce their deployment scenario requirements. Thus, a tunable approach has to allow for a specification of deployment scenario requirements on an abstract level that assumes as little expert knowledge on the security mechanisms as possible. The idea of separating requirements from their technical enforcement is the main idea of the policy-based management paradigm. In the following we provide an overview of the fundamentals of policy-based management and show how the tunable approaches we present in the remaining chapters of this thesis relate to it.

Policies enable a distinction between high-level objectives and the satisfaction of these objectives via low-level system implementations or configurations. This distinction enables a policy maker to achieve management objectives by specifying policies without having to address technical details regarding the implementation and configuration of low-level technical components [MS93, Slo94]. In particular, this comes with the advantage, that policy makers do not have to be experts regarding the low-level implementation of a system [Dam02]. Likewise, the entity that is responsible for operating or implementing the technical components is no longer obligated to address abstract management objectives but can configure or implement the system to satisfy deployment scenario specific policies. Applications of the policy-based management paradigm include access control [OAS13], network management applications [SLX01], and the application of service level agreements [BA07].

For the data outsourcing problem setting, the application of the policy-based management paradigm is shown in Figure 2.7. From the perspective of the risk manager, policies are tools to meet objectives, i.e., to mitigate risks with regard to higher level objectives. Thus, policies can be considered as deployment scenario requirements. From the perspective of the secure system, policies are the objectives that have to be enforced by implementing security mechanisms such as encryption. This separation of duties allows the risk manager to be oblivious to low-level security mechanisms and the outsourcing solution to be oblivious to high-level risk management objectives. In case the translation of policies to secure systems can be automated, complexities are further reduced as the risk manager can adapt policies and generate a new outsourcing solution on-the-fly once the deployment scenario changes without a deep understanding of the protection guarantees of the security mechanisms that are used to enforce policies.

In the following we summarize different aspects of policy-based management and highlight how they relate to the data outsourcing problem setting that is addressed in this thesis.

Policy specification: In a policy-based system, the policies are typically specified at the so-called *policy administration point* (PAP) and stored in a policy repository. To specify policies, a variety of domain specific languages exist that are tailored to the specific use-case. For instance, there exist a variety of languages to specify firewall policies and network routing policies [SLX01] as well as security [DDLS01] and access policies [PHB06]. For specific use-cases these languages have already been standardized. For instance, the eXtensible Access Control Markup Language (XACML) standardizes the representation and evaluation of access control policies [OAS13].

To our knowledge no standardized policy languages exist that are tailored to the secure data outsourcing problem settings that we addressed in this thesis. However, we make use of policy concepts that are widely used in the academic literature, including the Structured Query Language (SQL), so-called confidentiality constraints (see Section 3.5) and anonymity notions (see Section 3.6).

Policy evaluation: When applying policies, it can be distinguished between the *policy decision point* (PDP) and the *policy enforcement point* (PEP). The PDP determines which policies should be applied and makes a policy decision based on these policies. The PEP then enforces this policy decision. While the PDP depends on the policy framework that is used, the PEP depends on the technical foundations of the system. For instance, a policy decision based on the policy "user A may access server B" is "grant access for the current user" if the current user is user A. This policy decision can be enforced by an SSH server that acts as PEP and grants access to user A. In this example, a policy decision was made for a specific action, i.e., a user who requests access to a resource, in specific system configuration, i.e., the access control matrix of the system.

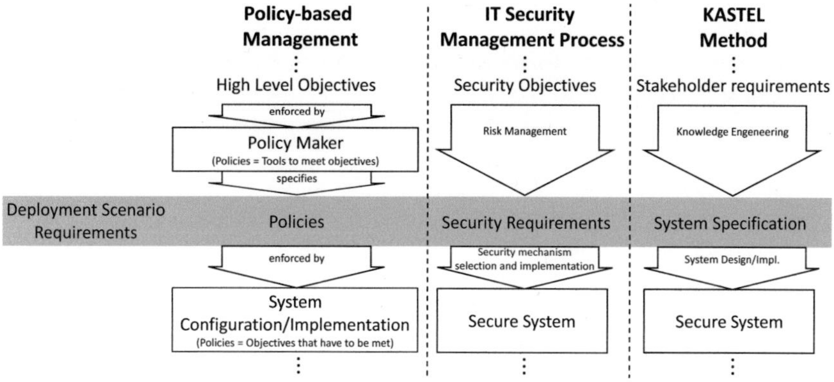

Figure 2.7: Policy-based management vs. the ISO 270xx IT security management process vs. the KASTEL method.

In other policy frameworks, the goal is not to generate a policy decision in a specific system configuration but rather to find a system configuration or design that satisfies all policies. For instance, *service level agreement* (SLA) policies each guarantee a specific throughput to a customer. The challenge is to find a system configuration, i.e., a mapping of resources on customers that does not violate any SLA.

The contributions presented in this thesis aim to find a system design or configuration that does not violate the policies, i.e., the deployment scenario requirements.

Policy refinement: Policies can be refined in a cascading way, i.e., more specific policies are derived from higher-level policies [MS93]. This reflects the hierarchical enterprise management structures where high-level business objectives are broken down into smaller objectives that are handed to departments. Compared to translating high-level policies directly into specific technical solutions, a policy refinement approach is modularized. In practical scenarios, it is not realistic to assume that high-level business objectives can be directly translated into specific technical solutions in an automated fashion without human interaction. This human interaction can take place within each policy refinement step.

With regards to the data outsourcing problem settings we address in this thesis, the risk manager may be subject to policies that are issued by the management of the organization and refines these policies to deployment scenario requirements. In the ISO 270xx process the higher-level policies are the security objectives, in the KASTEL method the higher-level policies are requirements that are specified by the stakeholder. Thus, the risk manager aims to satisfy the higher-level policies by specifying lower-level policies. In this thesis we address the final step of satisfying the policies via approaches that implement security mechanisms.

Policy conflict detection and resolution: In some cases policies can conflict with each other. For instance, the policy "user A may access file B" conflicts with "user A may not access file B". In some systems simple meta-policies can be used to resolve policy conflicts. For instance, a *deny-overrules* meta-policy expresses that the more restrictive policies are authoritative. However, in more complex settings policy conflict resolution is harder and potentially requires user supervision.

The approaches that we present in this thesis enable policy conflict detection and resolution. If the transformation method to match security mechanisms with deployment scenario requirements (policies) is as simple as mapping a set of requirements to a set of security mechanisms, policy conflict detection is simple. If no mapping exists for the deployment scenario requirements, the requirements are conflicting. To resolve the conflict, the risk manager has to determine whether another set of deployment scenario requirements would also satisfy the high-level security objectives or stakeholder requirements.

If the transformation method is more complex and many policies are defined, a policy conflict exists if no set of security mechanisms can be determined that satisfies the deployment scenario requirements. In many cases only a small subset of the specified policies are conflicting. However, from the perspective of the risk manager resolving the policy conflict is not trivial as it is not clear which policies are actually conflicting. A trial-and-error approach that alters specific policies to find a non-conflicting policy set is often not feasible due to the amount of candidate policies. The approaches we present in this thesis make use of mathematical programming to isolate conflicting policies and present them to the risk manager.

2.5.3 Fundamentals: Mathematical Optimization

To find a *feasible* combination of security mechanisms that satisfies given deployment scenario requirements, brute force approaches that traverse all possible combinations of security mechanisms are not practicable in most cases. Developing problem-specific algorithms from scratch to find feasible security mechanism sets is an option. However, in this thesis we reduce data outsourcing problems on optimization problem representations that have been well-researched in the operations research community and use existing optimized solvers to solve them. It's important to distinguish between the way an optimization problem is *modeled* and the way the modeled optimization problem is *solved*. In particular, the problem of finding security mechanism combinations that satisfy given deployment scenario requirements can be modeled as a satisfaction problem as well as an optimization problem.

Constraint Satisfaction problems: Constraint Satisfaction problems are about finding a variable allocation for a fixed set of variables that satisfies specified constraints. The problem of finding a set of security mechanisms that satisfies specified deployment scenario requirements can be mapped on a satisfaction problem. Techniques from the field of boolean satisfiability [Coo71] and the field of satisfiability modulo theories (SMT) [DMB11] or, more generally, constraint satisfaction [Tsa93] can be applied to solve the resulting problem. These techniques include backtracking [Apt03], constraint propagation [Apt03], and local search approaches [Apt03]. In case the problem is too large to be solved in feasible time by exact algorithms, heuristics can be used to approximate solutions to the problem. Depending on the heuristic, the produced solutions are not guaranteed to be optimal or feasible. While in most cases it is easy to verify that an approximated solution is feasible, it can happen that a heuristic finds no feasible solution even if one exists.

Techniques to solve satisfaction problems can be used to find a set of security mechanisms that satisfies specified deployment scenario requirements. In many practical scenarios optimized approaches are required that do not only satisfy the deployment scenario requirements but also optimize certain quality characteristics such as efficiency. A *feasible* solution that does not violate any deployment scenario requirement might not be an *optimal* one in practice. For instance, encrypting all attributes of a database before outsourcing it enforces the deployment scenario requirement that a specific attribute A may not be visible to the storage provider but only encrypting attribute A would suffice and be more efficient.

Constraint Optimization problems: Mathematical programming [BHM77] allows to specify a set of constraints that determine the feasible solutions and a target function that rates each feasible solution. The techniques that are used to solve the formulated optimization problems aim to find a feasible solution that either minimizes or maximizes the target function.

The complexity of solving mathematical programming problems depends on the properties of the optimization problem. On a high level, it can be distinguished between *convex optimization* [BV09] and *non-convex optimization* [Avr03]. Both the constraints and the target function of a convex programming problem have to be convex functions. In case of non-convex programming problems, the constraints and the target function may be arbitrary functions.

Techniques exist to solve convex programming problems efficiently. For instance, *linear programming problems* that require the constraints and the target function to be linear functions are a subclass of convex programming problems and can be efficiently solved by interior points methods [BV09] in polynomial time or by the simplex algorithm [NM65]. The simplex algorithm runs in exponential time in the worst case but has shown to be more efficient for practical problems than the interior points method.

Solving general non-convex programming problems is considered complex and cannot be achieved in polynomial time in many cases [BV04]. However, for some specific kinds of non-convex programming problems, techniques exist that showed good performance in practice. For instance, solving *integer linear programming* (ILP) problems were shown to be NP-hard but can be addressed by combining branch-and-bound [LD60] or branch-and-cut [PR91] techniques with efficient techniques to solve regular linear programming problems. Furthermore, to solve non-linear programming problems with differentiable constraints and target function, Karush-Kuhn-Tucker conditions [KT51] can be used to at least identify a subset of solutions that contains the optimal solution. If no approach exists that can solve a non-convex programming problem in appropriate time, heuristics with a better runtime that output close-to-optimal solutions can be used.

For the approaches presented in this thesis, binary decisions are necessary. For instance, a partial solution either contains the assertion "attribute A needs to be encrypted before being outsourced" or not. Unfortunately, this implies that the set of feasible solutions is not convex. The approaches presented in this thesis, make use of *Integer Linear Programming* [GN72] as they require binary decisions as well as integer variables/constraints and aim to optimize performance via a target function.

We abstract from concrete solving techniques by using generic ILP-solvers to solve the ILP problems that are formulated by our approaches.

2.6 Summary

We showed that in many problem settings trade-offs between quality characteristics have to be made. This is especially true for security characteristics. If a problem setting includes inherent trade-offs between quality characteristics, it is impossible to develop static approaches that satisfy all possible scenario requirements at once. We introduced the concept of tunability which allows to tune quality properties of an approach without having to re-design and re-implement the approach. Tunable approaches can tune trade-offs to match the requirements of the deployment scenario in which they should be applied. Thus, compared to static approaches they can be deployed in a multitude of scenarios even if the requirements of two such scenarios are conflicting. We illustrated this on the example of Facius, a tunable approach to integrate non web-based services with identity federations that is used in production within the bwIDM federation. Furthermore, we provided a methodology on how to develop tunable approaches and showed how findings of different research fields including policy-based management and operations research can be beneficially applied in the methodology. In the remaining chapters we show how the introduced methodology can be applied to build tunable approaches that address acknowledged problems in the domains of database, identity and credential outsourcing.

3
Database Outsourcing

The paradigm of outsourcing databases to external parties and evaluating database queries on the contained outsourced database records is subsumed under the term of *Database-as-a-Service* (DaaS). Outsourcing databases to cloud storage providers (CSPs) yields many benefits with regard to availability, scalability, and maintainability of the database as well as cost benefits (cf. Section 1.1). However, when data is outsourced to a CSP, data confidentiality and data integrity are at risk. For instance, a compromised CSP may try to sell the outsourced data to a competitor or manipulate the outsourced data to harm the business of the party that relies on the outsourced data. Possible attackers include both insiders such as malicious admins who work for the CSP and external attackers that compromise the infrastructure of the cloud provider. Note that the topic of database outsourcing is not limited to the cloud computing setting. Even inside of an organization databases can be "outsourced" in the sense that the database is operated by employees who do not have the security clearance to access the stored data [Sda14].

We apply the proposed methodology to build tunable approaches that address the DaaS problem setting in this chapter. We provide a detailed overview of the secure DaaS problem setting and define the scope of this thesis in Section 3.1. In Section 3.2, we specify the major research questions that arise and summarize previous work that relates to these research questions in Section 3.3. Following our proposed methodology, we first provide a comprehensive study of existing security mechanisms that preserve data confidentiality in Section 3.4 and assess them with regard to the level of protection they offer as well as with regard to the query functionality that they support. Which specific approaches should be applied in a deployment scenario depends on the scenario requirements with regard to confidentiality, the assumed attacker, and the queries that will be executed on the data. In Section 3.5, we propose the Securus framework which constitutes a tunable approach that allows to specify deployment scenario requirements in a domain specific language, automatically determines an optimized set of security mechanisms, and implements them in a software adapter. Besides the surveyed security mechanisms that enforce data confidentiality, security mechanisms that enforce data anonymity exist. We propose the Dividat approach that can be used to assess and optimize the efficiency of database anonymization techniques in Section 3.6. In Section 3.7 we discuss the differences between anonymized and confidential database outsourcing, highlight their individual benefits, and show how synergies can be leveraged. We summarize our contributions to the domain of database outsourcing and draw conclusions in Section 3.8.

Parts of the contributions presented in this chapter have been previously published in [KJ14, KJH14, KH14, KJH15, JKH12].

3.1 The Secure Database-as-a-Service Problem Setting

Securely outsourcing databases to external providers constitutes a complex problem setting that can manifest itself differently in particular depending on a) the relevant security characteristics which define "security" in a particular scenario, b) the assumed attackers, and c) which parties in the problem setting can be trusted. In the following, we provide an overview on the possible manifestations of the DaaS problem setting and clarify the scope of this thesis. We present the range of enforceable security characteristics (see Section 3.1.1), the range of attackers who can be assumed (see Section 3.1.2), and the possible trust anchors (see Section 3.1.3). Furthermore, we provide an exemplary deployment scenario that matches the scope of this thesis in Section 3.1.4.

3.1.1 Security Characteristics

Security in the DaaS problem setting can be subdivided into multiple security characteristics that can be required to be enforced. We list the most common security characteristics in the following.

Data confidentiality: The confidentiality of outsourced data is achieved if only authorized subjects can read the data's content. It can be distinguished between enforcing data confidentiality in a single-user setting in which only the data owner outsources and queries the database and multi-user settings in which multiple parties with different access rights to specific records outsource and query data [DdVF+05].

Anonymity of individuals: Anonymity is achieved if information contained in the outsourced data cannot be linked to individual persons by unauthorized subjects [FWCY10]. This objective can be achieved by applying approaches that enforce data confidentiality. However, approaches that enforce anonymity do not necessarily enforce confidentiality. Anonymization approaches can have advantages over confidentiality enforcement approaches with regard to efficiency.

Data integrity: The integrity of outsourced data is achieved if the outsourced database and query results cannot be manipulated by an unauthorized subject unnoticed. In particular, it can be a requirement to enforce that no database records can be manipulated, deleted, or inserted by unauthorized subjects without being noticed. Furthermore, it can be required that the results of database queries are correct, complete, and up-to-date [MNT06, XWYM07, WY14, WYX13].

Data availability: In many scenarios it is important that the outsourced data is indeed available and can be queried. This might not be the case if data was intentional deleted by an attacker or was never stored by the CSP to save costs [BJO09, JK07, SW08].

Scope of this thesis: We focus on enforcing the confidentiality and anonymity of individual records that are contained in outsourced relational data, i.e., data that contains *records* that consist of a fixed set of *attribute values*. We assume a single-user setting in which just one party outsources and queries the database. The other mentioned security characteristics are orthogonal to enforcing confidentiality and can be enforced by approaches that can be applied additionally to those presented in this chapter.

3.1.2 Attacker Models

In the DaaS problem setting it can be distinguished between honest-but-curious and malicious attackers. Furthermore, attackers can be assumed to be non-colluding.

Honest-but-curious attackers behave according to protocol and do not manipulate data or query results in any way but strive to gain confidential information from what they can observe. Examples for honest-but-curious attackers include CSPs that aim to extract data and either use it to achieve their own goals or sell it to third parties without the data owner noticing. CSPs that attack their users tend to be honest-but-curious attackers to prevent the user from noticing the attack by protocol violations or manipulated data. Once the data owner notices protocol violations or manipulated data, she will stop using the CSP service which conflicts with the primary business goals of most CSPs.

Malicious attackers do not behave according to protocol and can manipulate data and query results as well as pretend to store data when in fact they are not. They can do this to undermine security characteristics such as integrity and availability but also as an intermediary step to breach confidentiality. Examples for malicious attackers are external attackers that aim to harm the data owner. For instance, data can be manipulated so that all results that are derived from the data have to be considered tainted. In particular, if the data is used for research the credibility of the researchers that own and rely on the data can be undermined.

Non-colluding attackers: In some settings data is not outsourced to a single but to multiple CSPs. In the non-colluding attacker model it is assumed that a single attacker cannot compromise more than one CSP and does not collude with other attackers.

Scope of this thesis: In this chapter we assume that CSPs are honest-but-curious attackers and investigate how the data confidentiality of outsourced databases can be enforced even if CSPs are assumed to be compromised. In case multiple CSPs are used to outsource data we assume non-colluding attackers as the black bar in

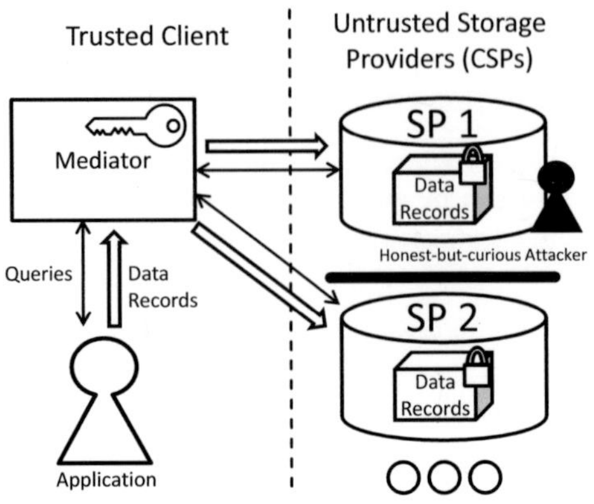

Figure 3.1: The Database-as-a-Service problem setting: Scope of this thesis.

Figure 3.1 indicates. We provide a more fine-grained attacker categorization of honest-but-curious attackers in Section 3.4 in order to assess and compare the protection guarantees of existing security mechanisms.

3.1.3 Trust Anchors

In this thesis, we use the term "trust" in a black and white fashion [VCDC11], i.e., an agent within a system is either trusted to perform specific actions or she is not. A trust anchor that enforces access control is necessary to enforce data security. For instance, access control can be enforced by a party that applies encryption. In this case, the party is trusted to correctly apply encryption. The trust anchor can either be established at the CSP (trusted cloud) or at the client application that outsources the data to the CSP (trusted client).

Trusted cloud approaches aim to achieve data security by establishing a secure environment at the CSP to store and query the data. This can be achieved by relying on trusted components such as secure co-processors [MT05] and trusted platform modules that have to be installed at the CSP [BS11]. Trusted cloud approaches are suited if the data has to be protected against third party attackers that compromised the CSP. If the outsourced data has to be protected against insiders with access to hard- and software, the protection offered by these approaches is significantly reduced. Furthermore, trusted cloud approaches require the CSP to install additional hardware features such as secure co-processors.

Trusted client approaches assume that the CSP cannot be trusted in providing a secure environment and pin the trust anchor at the client that outsources and accesses the data. For instance, data can be encrypted before being outsourced and decrypted upon retrieval by a trusted component that is executed by the client and acts as a *mediator* between the application that outsources data and the CSP.

Scope of this thesis: In this chapter we focus on a trusted client architecture that is shown in Figure 3.1. A user outsources data to one or more CSPs. While the application may access the data, the CSPs are considered compromised. Thus, data has to be protected before being outsourced to enforce data security. To achieve that, the application makes use of a mediator that applies security mechanisms such as encryption before outsourcing data and that also governs the execution of queries on the protected, outsourced data.

3.1.4 Exemplary Database Outsourcing Use-Case

As an example for secure database outsourcing consider the use-case of storing electronic health records [TBF14]. The records of hospital patients contain highly sensitive data. Storing those records in in-house databases requires in-house database experts. Especially for small hospitals this often constitutes a challenging requirement. To address this challenge, electronic health records can be outsourced to external CSPs. However, in many cases it is not clear whether outsourcing the data undermines confidentiality. For instance, this can be the case if the CSP is not sufficiently protected against third party attackers or employees of the external CSP are honest-but-curious attackers. Note that the security problems that arise with insider threats also exist for in-house health record databases [Sda14]. The contributions presented in this chapter are not limited to the outsourcing scenario, but can provide protection against insider attacks as well.

An attacker that compromised the cloud provider can monitor queries. The attacker is interested in eavesdropping patient data over extended time periods and remaining undetected. For instance, the goal of the attacker can be to sell the eavesdropped patient data or to use it to perform social engineering attacks. As manipulations on the data increase the risk of being detected, the attacker can be assumed not to manipulate any data. Data confidentiality and the anonymity of patients have to be enforced against such an attacker. This can be done by providing secret keys to the stakeholders who are allowed to access the electronic health records database and let them access the encrypted database via a mediator as shown in Figure 3.1.

Consider the exemplary database shown in Table 3.1. To protect the privacy of the patients it is important to enforce the anonymity of patients in the sense that it must not be possible to map illnesses to patients. Other attribute values such as the social security numbers or x-ray pictures may be considered confidential on their own and have to be protected.

internal id	name	social sec. number	x-ray pic.	illness	room number	...
1	Adam	12345	□	Headache	102	...
2	Bob	25123	□	Flu	212	...
3	Carol	35613	□	Cancer	112	...
...	...	...	...	...	...	...

Table 3.1: Examplary database: electronic health records.

Most importantly, notice that some information does *not* have to be protected. For instance, the internal id of a patient has no meaning outside of the database. Furthermore, the room numbers of the hospital may not be considered confidential. The mapping of room numbers to patients on the other hand can be considered confidential. As we show in the remainder of this chapter, approaches that can be tuned to satisfy specific deployment scenario requirements can improve the efficiency of evaluating queries on the database by leveraging the fact that some outsourced data does not require protection.

3.2 Specific Research Questions

In the DaaS domain, the security characteristics confidentiality and anonymity overlap in the sense that it is possible to achieve anonymity by providing confidentiality[1]. However, the approaches that were proposed to enforce confidentiality and anonymity differ significantly. Confidentiality enforcing security mechanisms aim to hide as much information as possible from attackers. Anonymization approaches aim to hide the relationship between individuals and sensitive information while at the same time exposing as much information as possible that is not related to individuals.

In the following, we distinguish between confidential Database-as-a-Service and anonymized Database-as-a-Service. While the confidentiality preserving DaaS approaches can be used to achieve anonymity by enforcing confidentiality, the anonymized DaaS approaches aim to achieve anonymity via traditional anonymization techniques that aim to expose as much information as possible without violating the anonymity requirements. Confidential and anonymized DaaS approaches have advantages and disadvantages over the other. Anonymized data constitutes noisy plaintext data based on which a CSP can evaluate complex queries. The anonymized data is noisy in the sense that some information is removed from it to preserve the anonymity of individuals. Confidential DaaS security mechanisms apply cryptographic methods and outsource ciphertext values instead of plaintext values in most cases. Based on these ciphertext values it is possible to evaluate simple query conditions more efficiently than on noisy plaintext data. In both problem settings an inherent trade-off exists between confidentiality/anonymity and efficiency with regard to query latency, transmission, and storage overhead.

We introduce the research questions that we address for the confidential Database-as-a-Service setting in Section 3.2.1 and for the anonymized Database-as-a-Service setting in Section 3.2.2.

3.2.1 Confidential Database-as-a-Service

Various security mechanisms exist that can be used to protect the confidentiality of outsourced data and still allow to execute query workloads efficiently, i.e., query the data in an efficient way. However, these security mechanisms differ with regard to

[1] This is true from a technical perspective. From a legal perspective, whether mechanisms such as encryption are sufficient to achieve anonymity is a controversial subject.

the queries they support, their ability to provide protection against different attacker capabilities, and their efficiency. Thus, a single security mechanism usually does not support the efficient execution of queries that are required to be executed in a given scenario (query requirements). Multiple security mechanisms have to be combined to satisfy the requirements of most deployment scenarios. Furthermore, usually not every part of the data is considered confidential and requires the same level of protection (confidentiality requirements). Efficiency can be optimized by applying only the security mechanisms that are necessary to protect confidential parts of the data. Choosing an appropriate combination of security mechanisms that satisfies the confidentiality requirements and optimizes efficiency is hard. This leads to the following questions that need to be answered to develop an approach for a specific deployment scenario:

- Which combination of security mechanisms satisfies the confidentiality requirements of the deployment scenario?
- Which suitable combination of security mechanisms is optimal in terms of efficiency?

To build a tunable approach that does not have to be re-designed and re-implemented by a cryptography expert for different sets of confidentiality and query requirements, it is necessary to develop an automated transformation method that transforms requirements with regard to efficiency and confidentiality to a suitable and efficiency-optimized set of security mechanisms. The key challenge for such a transformation method is to optimize the workload efficiency without violating the confidentiality requirements. This leads to the following research question that is addressed in this chapter with regard to confidential DaaS:

How can an efficiency-optimized database outsourcing solution that satisfies confidentiality requirements be derived for a given query workload?

3.2.2 Anonymized Database-as-a-Service

Existing anonymization approaches apply noise to the data but otherwise leave the data in plaintext. Outsourcing the noisy data on its own is no option in most DaaS settings due to the fact that most applications are not designed for noisy query results and require exact query results. Inexact query results can be avoided by outsourcing securely encrypted records and only anonymizing index tables that can be used by CSPs to participate in query evaluation. The CSP can evaluate queries on the noisy records of the index table and return the encrypted records that are linked to the matching noisy index table records to the mediator. The original records that do not contain noise can be restored at the mediator by decrypting the encrypted records of the result. The noisy index tables can lead to the transmission of encrypted records to the mediator that actually do not match the query. After decryption, these records can be filtered out by the mediator by checking if they match the query conditions. Compared to plaintext index tables, anonymized index tables reduce efficiency as records are unnecessarily transmitted. Thus, there exists a trade-off between efficiency and noise that is applied for anonymization.

For a given set of data records, existing anonymization techniques can produce different anonymized index tables that all satisfy the same set of anonymity requirements but differ with regard to their efficiency. An approach that allows to tune the trade-off between anonymity and efficiency has to determine an efficiency-optimal anonymized index table that satisfies the given anonymity requirements. To optimize the efficiency of anonymized indexes, the efficiency impact of approaches that are used to build anonymized indexes and their parameterization has to be understood and modeled. This leads to the following research question that is addressed in this chapter regarding anonymized DaaS:

How can efficiency models for anonymized database indexes be built?

3.3 Related Work

Confidential Database-as-a-Service

To enable confidential database outsourcing, frameworks exist that integrate the confidentiality preserving indexing approaches which we survey in Section 3.4. We propose Securus which constitutes such a framework in Section 3.5. In the following we provide an overview of comparable frameworks, highlight the differences and show which frameworks are suited for which deployment scenario.

Onion-encryption approaches: Popa et al. proposed the CryptDB framework [PRZB11, PZB11] which encrypts the entire database and makes use of so-called onion-encryption to loosen the confidentiality guarantees adaptively once queries are executed. In simple terms, onion-encryption works as follows. To enable the evaluation of range queries, each attribute value is encrypted by an order-preserving encryption scheme (first encryption onion). The resulting order-preserving ciphertexts are encrypted with a probabilistic encryption scheme (second encryption onion). The resulting indistinguishable ciphertexts cannot be used by an attacker to learn anything about the data's content and can be outsourced to the CSP. Once a query that contains a range condition on an attribute has to be executed, the CSP notifies the mediator that it is not able to evaluate the query based on the ciphertexts of the probabilistic encryption scheme. The mediator then sends the decryption key for the second encryption onion to CSP. The CSP can use the key to decrypt the probabilistic ciphertexts and get access to the contained order-preserving ciphertexts. Thus, the CSP learns the order of the attribute values and is able to evaluate the range query.

As long as no queries are executed, the CSP only has access to indistinguishable ciphertexts of probabilistic encryption schemes. Once queries are executed, the confidentiality guarantees are considerably weakened as encryption onions that prevent an efficient query execution are decrypted. In this regard, onion-encryption can be seen as opportunistic encryption and confidentiality is not strictly enforced. For instance, to execute the TPC-H benchmark [Tra14] 32.80% of the outsourced attributes have to be decrypted to order-preserving ciphertexts [GHH+14]. In the worst case, the executed queries result in the entire database to be decrypted to plaintext. Thus, unlike Securus, onion-encryption approaches like CryptDB are not suited to enforce strict confidentiality requirements.

Kerschbaum et al. [KHK+13] addressed this problem by adding a policy layer to onion-encryption that allows the mediator to check whether onions can be decrypted without violation of the specified confidentiality requirements. For instance, a user can specify that the onions of an attribute *a* must not be decrypted further than order-preserving ciphertexts. To satisfy this requirement, the mediator will never release the encryption keys for the order-preserving ciphertexts of attribute *a*. If an onion that has to be decrypted to evaluate a query condition must not be decrypted, the condition cannot be evaluated by the CSP and false-positive records that do not match the query have to be transmitted the mediator. The mediator then decrypts the values and evaluates the remaining query condition. In Section 3.5.6 we show that the efficiency overhead for transmitting false-positive records can be substantial and depends on the outsourced data's structure.

Compared to Kerschbaum et al. and Popa et al.'s approaches, the efficiency of outsourcing solutions based on Securus is higher due to the fact that 1) Securus only protects what is necessary and 2) Securus requires that queries are specified up front and can leverage this information to optimize efficiency.

1) Compared to Kerschbaum et al. and Popa et al.'s approaches, Securus only applies CPIs if it is necessary to satisfy the specified confidentiality requirements. Even encryption schemes that are considered efficient by CryptDB induce a certain overhead compared to outsourcing plaintext data[2]. Securus avoids to protect attributes unnecessarily.

2) Securus leverages that future queries are specified up front to optimize the set of CPIs that have to be applied for efficiency. For instance, consider a scenario in which either attribute *a* or attribute *b* must not be revealable by the CSP. Furthermore, a single query with a condition on attribute *a* is evaluated before a large set of queries with conditions on attribute *b*. Onion encryption approaches would decrypt attribute *a* as it is queried first. Securus avoids that attribute *a* is decrypted for maximum efficiency and favors decrypting attribute *b* instead. Furthermore, Securus avoids unexpected efficiency overheads induced by transmitting false-positive records as it is checked beforehand whether the specified queries can be evaluated at the CSPs.

As we show in Section 3.5.6 Securus can achieve more efficient solutions than adaptive onion-encryption approaches. However, Securus requires the user to a-priori specify the future query workload. In future work Securus can be combined with onion-encryption to also support further queries beyond those that were specified in the policy profile and for which the mediator is optimized. Thus, queries that were not known to be relevant at the time the policy profile was specified can be executed in a best-effort way.

[2] This is especially true for insert and update operations, as each onion has to be updated even if the encryption onion is not required by any future query.

Fragmentation-based approaches: The concept of splitting up the attribute values of a record and storing each attribute in a separate data table [ABG+05] can be used to protect attribute value combinations rather than the attribute values themselves. Achenbach et al. proposed the Mimosecco framework [AGH11] which applies fragmentation to protect an entire database. Their approach protects every possible attribute value combination. The values of each attribute are stored in a separate table in plaintext along with encrypted record IDs of the records that they are part of. As the record IDs are encrypted, the CSP is not able to link the attribute values. To evaluate a query, the CSP can evaluate each query condition separately on the table that contains the attribute that is required to evaluate the condition and return the encrypted record IDs of the matching values to the mediator. The mediator decrypts the record IDs, intersects the results of different conditions and retrieves the according encrypted records from the CSP.

Huber et al. proposed the Cumulus4j [HGSB13] framework which also applies fragmentation to protect an entire database. If attribute values of a single attribute value have to be protected, the attribute is not indexed and therefore not outsourced in plaintext. Besides providing a framework, they formalized a security notion that Cumulus4j can achieve.

The problem that Mimosecco and Cumulus4j share is that a lot of encrypted record IDs are transmitted unnecessarily which substantially affects query execution efficiency. For instance, given that 50% of the outsourced records map to a male person, 50% of all encrypted IDs have to be transmitted to the mediator to evaluate the conditions `name=john` and `gender=m` even if only one record would satisfy both conditions at once. Thus, many of the transmitted record IDs constitute false-positives, i.e., they map to records that do not match all of the query conditions.

Vimercati et al. [DCdVFJ+14, For11, DFJ+13b] address this problem by using fragmentation only to protect the attribute combinations that are specified as confidential by the user. Thus, attributes can occur together in plaintext in one table and the number of transmitted records can be minimized. Based on a user-specified query workload their approach is capable of finding a fragmentation strategy that satisfies the confidentiality requirements and minimize the number of transmitted false-positive records. However, we show in Section 3.5.6 that the efficiency overhead for transmitting false-positive records can still be substantial.

Compared to the introduced fragmentation approaches, Securus makes use of fragmentation but enforces that each query is evaluated based on a single fragment to avoid the transmission of false-positives. Furthermore, Securus applies CPIs to avoid the need for data fragmentation. This also allows Securus to protect single attributes while still being able to evaluate query conditions on it.

Recommendations: Each approach has advantages over the others in some respects. Which approach is suited best for a given deployment scenario depends on whether confidentiality has to be guaranteed as well as whether precise requirements are known with regard to data confidentiality and queries that will be executed in the future. Our recommendations on which approach to use are summarized in the following:

- Unknown query and confidentiality requirements, opportunistic encryption sufficient: use CryptDB [PRZB11, PZB11].
- Unknown query and confidentiality requirements, guaranteed security properties required: use Mimosecco/Cumulus4j [AGH11, HGSB13].
- Unknown query and known confidentiality requirements, guaranteed security properties required: use Kerschbaum et al.'s approach [KHK+13] or combine it with Securus for improved efficiency of queries that are known when the policy profile is specified.
- Known query and confidentiality requirements, guaranteed security properties required: use Securus [JKH12, KJ14, KJH14] to leverage the knowledge and improve efficiency as much as possible.

The main reason why Securus is better suited than other approaches when query and confidentiality requirements are known is increased efficiency. We provide a fine grained comparison between the approaches' efficiency in Section 3.5.6.

Other related frameworks: Approaches were proposed that allow to evaluate MapReduce and PigLatin queries on encrypted data [TLMM13, JSSSE14, SSSE14]. They address the execution of analytical data-processing query workloads. In comparison, the Securus approach we present in this chapter focusses transactional data-processing based on relational databases. Furthermore, they aim to encrypt everything instead of selectively applying security mechanisms to satisfy the deployment scenario requirements. This results in efficiency overheads that could be avoided. In this sense, the approaches cannot be considered tunable. Furthermore, query conditions that cannot be evaluated based on the outsourced ciphertexts are evaluated by the mediator which can result in unexpected efficiency overheads.

To our knowledge, unlike Securus, other related approaches do not allow to specify the attackers' capabilities. Thus, the weakest security mechanisms they apply limit the applicability against specific attackers and only "weak" attackers can be addressed. The Securus approach allows to specify attacker capabilities. Thus, it is possible to make selective use of weak security mechanisms to increase efficiency if they can be considered sufficient to protect against the assumed attacker. Securus can be tuned to provide protection against stronger attackers by specifying stronger attacker capabilities. In this case, Securus only makes use of stronger security mechanisms.

Anonymized Database-as-a-Service

Anonymity notions: A variety of notions exist that capture the anonymity level of the individuals that are contained in a set of data records [FWCY10]. The anonymity notions that are relevant to understand the content of this thesis distinguish between sensitive attributes and attributes that can be used to re-identify an individual in the data. The set of attributes that can be used to map an individual to its data record is called quasi identifier (QID)[3]. Ideally, an attacker should not be able to map sensitive attribute values to an individual. By definition, the QID can be used to map individuals on their data records. To prevent the mapping between individuals and sensitive attribute values, it must not be possible to map an allocation of the QID to a specific sensitive attribute value.

The k-anonymity [Swe02b] notion demands that each QID allocation occurs at least k times in a set of records. Thus, it is not possible to determine which of the k records maps to the record of an individual *i*. However, if all k records map to the same sensitive value, an attacker that is interested in the sensitive attribute value of the individual *i* does not need to find the precise record that maps to *i*, as all candidates map to the same value. Thus, the attacker is able to reveal sensitive information on *i* regardless of whether the outsourced records satisfy k-anonymity.

To remedy this shortcoming of k-anonymity, the notion of ℓ-diversity [MKGV07] was proposed. Intuitively, a set of records is considered ℓ-diverse if each QID allocation maps to at least ℓ distinct, well-represented sensitive values. As it is guaranteed that each QID allocation maps to at least ℓ distinct sensitive values, the introduced attack against k-anonymity no longer works.

Sophisticated attacks against ℓ-diversity still exist which led to proposals of stricter anonymity notions. Among them are t-closeness [LLV07], m-invariance [XT07] and differential privacy [Dwo06]. It has been shown that satisfying stricter anonymity notions becomes increasingly complex and costly. As we exemplarily apply our contribution Dividat on approaches that achieve ℓ-diversity, we do not present all anonymity notions in this thesis. However, Dividat is not limited to approaches that meet ℓ-diversity requirements but can also be applied to approaches that satisfy other anonymity notions.

Anonymization techniques: Various techniques exist to satisfy specific notions of anonymity. To give an impression of the range of anonymization techniques we provide an overview of the anonymization technique categories.

Generalization [LDR05, FWY07] enforces anonymity by censoring the values of the QID attributes in such a way that each QID value combination occurs more often and it is thus harder to map an individual to a record. For instance, numerical attribute values can be generalized by suppressing certain number of digits so that "12345" and "12389" both become "123**". For categorical attributes, a taxonomy can be used to generalize attribute values. For instance, "Painter" and "Musician" can both be generalized to "Artist".

Anatomization [XT06] enforces anonymity by hiding the relationship between QID attributes and sensitive attributes. This can be achieved by splitting up the QID

[3] A QID is formally defined as follows [Swe02b]: Given a population of entities U that are contained in the data and the set of total entities U', an entity-specific table $T(A_1, ..., A_n)$, $f_c : U \rightarrow T$ and $f_g : T \rightarrow U'$, where $U \subseteq U'$. A quasi-identifier of T, written QID, is a set of attributes $A_i, ..., A_j \subseteq \{A_1, ..., A_n\}$ where: $\exists p_i \in U$ such that $f_g(f_c(p_i)[QID]) = p_i$. "Intuitively, f_c maps an entity p_i on its according record in table T. QID is a quasi-identifier if there exists a mapping f_g that can map the record $f_c(p_i)$ back to the original entity p_i based on the attributes contained in QID." [KH14]

and the sensitive attributes on multiple tables [DFJ+13a]. Thus, an attacker may be able to map an individual on its QID allocation in the QID table but is unable to map a sensitive value to it.

Suppression [WFY07, LDR06] enforces anonymity by not outsourcing certain attributes at all. In the extreme case, no QID attributes are outsourced. Thus, the attacker is not able to map an individual to its database record and, therefore, its sensitive attribute values.

Perturbation deliberately adds noise to attribute values that are contained in the table. For instance, this can be achieved by swapping attribute values of different records [Rei84] or by outsourcing generated data that has the same statistical characteristics as the original data [AY08]. Using perturbation in DaaS scenarios is problematic as the results of a query cannot be assumed to be complete. For instance, if attribute values are swapped between records, a query selecting all records that contain the swapped attribute value returns a record that does not fit the query (false-positive) and does not return the record that actually fits the query (false-negative). Whereas false-positives can be filtered out by the mediator, it is not possible to recognize false-negatives records that were not transmitted to the mediator.

Counterfeit Records [XT07] can be inserted into a data table so that the required anonymity notion is satisfied. This method can add a lot of artificial information to the original data which makes it undesirable for data-publishing. However, in DaaS scenarios, the anonymized table can be extended by encrypted statements whether a record is counterfeit or not. Thus, it is possible for the mediator to determine counterfeit records from real ones while the CSP cannot distinguish them.

Anonymization techniques for continuously updated data: Traditional anonymization techniques focus on anonymizing a set of data that is not modified in the future. In most deployment scenarios, the outsourced data records are modified. However, applying updates to already published anonymized data can undermine the achieved anonymity notion. More recent approaches address this problem and investigate how anonymized data can be updated without undermining specific anonymity notions. In particular, approaches were proposed to achieve the notions of m-invariance [HBN11][XT07] and k-anonymity [FWFP08][ZHP+09][PXW+07] in presence of updates on the data. The approaches apply the strategy to suppress updates until a large enough number of updates can be performed in one batch. Each batch of updates can be anonymized, so that the anonymity of the unmodified records and the updated records is not harmed.

Anonymized database-as-a-service approaches: In most DaaS deployment scenarios, databases are frequently updated and it is required that executed queries return correct and up-to-date results. Anonymization techniques for continuously updated data aggregate updates until they can be applied to the outsourced data without harming anonymity. Thus, it is not guaranteed that each update is reflected in the outsourced data right away and query results may be incomplete, i.e., records that would actually match a query are not returned as an already issued update was not yet applied to the outsourced data. To achieve the completeness of query results, several approaches were proposed for the DaaS setting.

Nergiz et al. [NCM13, NC11] leverage findings from the continuous data publishing community to enable ℓ-diversified database outsourcing. They divide the outsourced records into groups of at least ℓ records and assign a group ID to each group. Then they anatomize the data by storing the QID attributes and the sensitive attributes along with the assigned group IDs in two separate tables at the CSP. By doing that, the binding between the QID part of the records within a group and their sensitive attribute values are lost. Each QID allocation maps to ℓ sensitive attributes and the ℓ-diversity requirements are met. Newly inserted records or updates on existing records are stored in separate fully encrypted tables until they can be committed to the anatomized data without violating the ℓ-diversity notion. These tables are downloaded and decrypted by the mediator for each query so that query result completeness can be guaranteed.

Vimercati et al. [DFJ+13a, DCdVFLS11] propose a similar approach that also builds on anatomization and supports attributes that are both sensitive and contained in the QID. They show that the anonymity notions that can be achieved by their approaches are comparable to ℓ-diversity.

The Dividat framework we present in Section 3.6 is not a new anonymization technique for anonymized database outsourcing but a framework to develop models to measure and optimize the efficiency of existing anonymized indexing approaches. Dividat can be applied on existing database anonymization approaches. In Section 3.6.2 we explain the approaches we exemplarily apply Dividat on in more detail. To our knowledge, we are the first to propose a framework to capture and optimize the efficiency overheads that are incurred by anonymized indexes in DaaS scenarios.

Measurement of the utility of an anonymized database: In the privacy-preserving data publishing community the overhead that is induced by anonymization approaches was measured based on general purpose information metrics [Iye02, Sam01, Swe02a, Swe02b]. For instance, the quotient of the number of different attribute values before and after applying generalization makes an assertion on the information loss with regard to data granularity. These general purpose information metrics only covered the information loss that is incurred by anonymization techniques such as generalization without making assertions on the impact on the actual use-case for which the data was published. Other approaches measured the overhead of anonymizing data by checking the quality of classifiers that were built based on the data [FWY07, LDR06, Iye02]. In the privacy-preserving data publishing problem setting, building classifiers is a common use-case which makes classifier quality a meaningful metric. In the database outsourcing problem setting the objective is to evaluate queries as efficiently as possible. Dividat can be used to determine efficiency models that assess different anonymized index tables with regard to meaningful database outsourcing metrics such as the induced query latency, transmission, and storage overhead.

Database index optimization: Similarly to Dividat, efficiency modeling methodologies were proposed for the physical design and query tuning of non-anonymized databases. Physical design tuning [CN07, ACN06, BC05] has the objective of finding a set of indexes that minimizes the execution costs of a query workload W while not violating a storage budget B. Query tuning [CCG+99, CN97] aims to find the most efficient method to evaluate a query on a database. This includes choosing an appropriate index.

Physical design and query tuning approaches are based on the assumption that the query can be fully executed within the database management system and the result of the query that is passed to the relying application can be considered final. This is not the case in anonymized database outsourcing scenarios, as the query results are noisy and encrypted when they leave the database management system of the CSP and have to be post-processed by the mediator before the final, de-anonymized query result can be passed to the relying application. Thus, traditional physical design and query tuning approaches are insufficient to optimize query execution and index creation in the anonymized database setting. In particular, they do not consider the network link between the mediator and the CSP as well as the transmission of false-positive records that leads to increased query latency and transmission overhead. However, Dividat makes use of these existing techniques in the sense that it relies on traditional database management systems that perform query tuning.

3.4 Study of Confidentiality Preserving Indexing Approaches

Enabling an efficient query evaluation on outsourced data and protecting data confidentiality at the same time constitutes the primary challenge in the confidential DaaS problem setting. Addressing this challenge constitutes an active field of research. Various security mechanisms already exist that address parts of this confidential database outsourcing challenge and can be used to enforce the confidentiality of attribute values and reduce query efficiency overheads by allowing the CSP to participate in the evaluation of queries. In this section we identify and assess these security mechanisms. In the following we denote such security mechanisms as *confidentiality enforcing indexing approaches* (CPIs). Existing CPIs are heterogeneous in the sense that they differ in the level of protection they provide against attackers with specific capabilities and support different types of queries. Thus, CPIs that are suited for one deployment scenario might not satisfy the confidentiality and efficiency requirements of another deployment scenario. Determining which CPIs are suitable for a given deployment scenario is not easy as expert knowledge on the CPIs and possible attack vectors is required. In this section we assess existing CPI approaches and categorize them according to their ability to enable an efficient query execution and their protection guarantees against various attackers. The resulting CPI catalog constitutes a mapping of CPIs to deployment scenario requirements.

We first provide a taxonomy in Section 3.4.1 that captures deployment scenario requirements and can be used to categorize CPIs according to their supported query functionality and their protection guarantees against different attackers. In Section 3.4.2 we provide a methodology that can be used to assess and categorize CPIs with regard to the taxonomy. We apply this methodology in Section 3.4.3 on a wide range of existing CPIs to build a catalog that can be used to determine which CPI is suitable to satisfy given deployment scenario requirements. Furthermore, we assess the efficiency of the investigated CPIs in Section 3.4.4. In Section 3.4.5, we draw key conclusions and provide recommendations on which CPIs induce the least efficiency overhead for each attacker type and required query functionality.

3.4.1 CPI Taxonomy

In the following we provide a CPI taxonomy that can be used to assess CPIs according to their supported query functionality, i.e., the types of queries that are supported by the CPI, the guaranteed protection level, i.e., the degree to which confidentiality is preserved, and the attacker models against which a CPI can give those guarantees. The taxonomy can be used to express the properties of CPIs as well as deployment scenario requirements. Thus, a CPI categorization according to the taxonomy can be used to map deployment scenario requirements on CPIs that satisfy them.

Query Functionality

The main focus of CPIs is to protect the outsourced data's confidentiality and still allow for an efficient query execution. Queries are used to retrieve certain records or aggregated data values from the outsourced data. Each CPI focuses on allowing for an efficient execution of queries that contain specific query components, e.g., conditions that a record has to satisfy to be included in the query's results. For the CPI study we focus on the following subset of query components.

- *Equality selections (ES)* can be used to query for records that contain a certain attribute value.

  Example: `SELECT ... WHERE pseudonym='Adam'`

- *Range selections (RS)* can be used to query for records that contain an attribute value within a certain range.

  Example: `SELECT ... WHERE salary<30000`

- *Like selections (LS)* can be used to query for records that contain an attribute value that are similar to a given expression.

  Example: `SELECT ... WHERE pseudonym LIKE 'Ad'`

- *Aggregations (AG)* can be used to query for aggregated attribute values of multiple records.

  Example: `SELECT SUM(salary) ...`

The list of query components must not be understood as exhaustive but as a foundation for our CPI study. We show in Section 3.5.3 how the results of the study can be easily used to determine the suitability of CPIs to support more complex SQL query components such as "GROUP BY".

Aside from requirements regarding the querying of the outsourced data, many deployment scenarios require to modify the outsourced data. In particular, *insertions*, *updates*, and *deletions* of outsourced data records have to be performed. CPIs also differ with regard to whether they support those data modification operations.

ID	pseudonym	rating	position	salary	age
1	Adam	1	1	20000	23
2	Bob	2	2	20000	25
3	Carol	3	2	24000	25
4	Dan	4	2	20000	30
5	Eve	5	3	26000	30
6	Adam	6	2	40000	30

(a) Plaintext data

ID	pseudonym	rating	position	salary
α	Adam	345	θ	ι
β	Bob	452	η	κ
γ	Carol	632	η	λ
δ	Dan	742	η	μ
ϵ	Eve	893	ζ	ν
ϕ	Adam	897	η	ξ

ID	age
π	23
ρ	25
σ	25
τ	30
υ	30
ϕ	30

rating: order-preserving ciphertexts
position: distinguishable ciphertexts
salary and ID: indistinguishable ciphertexts

(b) Data that is outsourced to the CSP

Table 3.2: Running example: plaintext and outsourced database.

Protection Levels

CPIs focus on preserving the confidentiality of outsourced data records. The conception of whether the confidentiality of a single record can be considered protected depends on the deployment scenario requirements. For instance, some deployment scenarios only require to protect the confidentiality of specific attribute values that are contained in each record. In other deployment scenarios, attribute value combinations have to be protected, i.e., specific attributes must not be readable by an attacker together. Note that attribute combinations can be protected by protecting single attributes. In the CPI study we will focus on the protection of single attributes. We show in Section 3.5.2 how CPIs can also be used to protect attribute value combinations.

> *Example:* Consider that the confidentiality of a record in Table 3.2a can be considered protected if it is not possible to map a *pseudonym* to a *salary*. This requirement is enforced in Table 3.2b by only storing ciphertexts of *salary* at the CSP. Thus, the confidentiality of the *salary* attribute is protected and it is not possible to map a *pseudonym* to a *salary*.

CPIs can be categorized with regard to their conception of *record protection*, i.e., when they consider *all* the outsourced records to be sufficiently protected.

- *Computational record protection (CRP)*: A CPI guarantees computational record protection if it prevents a computationally bounded attacker that compromised the CSP from revealing confidential attribute values which she would not have been able to without having compromised the CSP.

- *Probabilistic record protection (PRP)*: A CPI guarantees probabilistic record protection if an attacker might be able to infer confidential attribute values of a small set of records or the attacker can narrow down the set of possible confidential attribute values of specific records. For the sake of simplicity, the definition of probabilistic record protection is intentionally kept vague. The PRP protection level can be subdivided further in future work in order to enable a more fine-grained specification of deployment scenario requirements. Securus only relies on CPIs that offer computational record protection. It can be extended in future work to leverage CPIs that provide probabilistic record protection.

  Example: In Table 3.2b, computational record protection is guaranteed for the attribute combination of *pseudonym* and *salary* as an attacker learns nothing about Adam's salary without the decryption key for the *salary* ciphertexts which are indistinguishable to her.

  For the attribute combination of *pseudonym* and *age* only probabilistic record protection can be guaranteed. An attacker is not able to determine Adam's age by joining the tables, as the ID's are encrypted. However, the attacker learned that Adam is either 23, 25 or 30 years old as no other option exists in the table that contains *age*.

Assumptions on Attackers

In the confidential DaaS setting, we assume *honest-but-curious attackers*, i.e., passive attackers that aim to eavesdrop data but do not maliciously manipulate any data. We provide a taxonomy for such attackers in the database outsourcing setting. We show in Section 3.4.2 how to determine against which attacker models of the taxonomy a CPI can enforce confidentiality. An overview of our attacker taxonomy is shown in Figure 3.2. We differentiate between the *eavesdropping capabilities* and the *background knowledge* of an attacker.

Attackers can be categorized according to their *eavesdropping capabilities*, i.e., their abilities to observe the outsourced database:

- *Access to the outsourced data (D)*: A D-Attacker can read all of the data that is outsourced to the CSP.

  Example: An attacker that compromises a CSP, downloads the data, is discovered and locked out from the CSP again constitutes a D-attacker.

- *Ability to eavesdrop data modifications (M)*: An M-attacker is able to observe changes that are applied to the outsourced data. We assume that an M-attacker is also able to reconstruct the outsourced data based on observed modifications and thus also has the abilities of a D-attacker.

  Example: A curious database admin who can read the database's content but is not able to install any software on the system to monitor incoming queries can observe modifications on the data by comparing the state of the database at different times. Thus, the attacker would be considered an M-attacker.

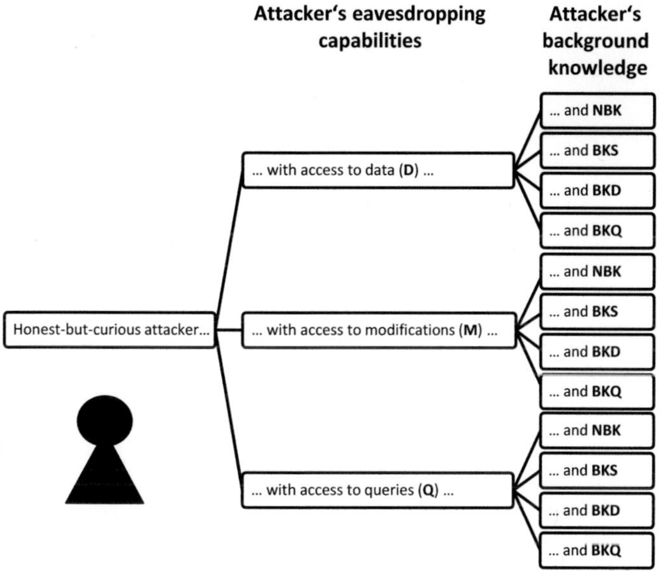

NBK: No background knowledge
BKS: Background knowledge of the data's schema
BKD: Background knowledge of the data's content
BKQ: Background knowledge of the data's content and queries

Figure 3.2: Attacker taxonomy [KJH15].

- *Ability to eavesdrop queries on the data (Q)*: A Q-attacker is able to observe queries that are executed on the data. In particular, a Q-attacker also has the eavesdropping capabilities of an M- and a D-attacker as INSERT, UPDATE, and DELETE operations are triggered by queries that can be observed.

  Example: A curious server admin that has root privileges on the database server is able to inspect the database's content and can install monitoring tools to observe queries that are executed on the data. Therefore, such an attacker is considered a Q-attacker.

Most CPIs do *not* preserve the confidentiality of attributes in general, i.e., information such as the frequency distribution of attribute values may still leak. Whether such leaked information harms the confidentiality of individual records strongly depends on the background knowledge of the attacker. Attackers can be categorized according to their *background knowledge*, i.e., the knowledge they possess aside from their observations of the outsourced database:

- *No background knowledge (NBK)*: If an attacker has no background knowledge on the data we call her NBK-attacker. To an NBK-attacker the eavesdropped data is the only source of information.

Example: If an attribute such as a randomly generated ID is only used in-house and the attacker is not assumed to be an employee, it can be assumed that the attacker has no background knowledge on this attribute.

- *Background knowledge on the data's scheme (BKS)*: If an attacker has background knowledge on the data's scheme we call her BKS-attacker. In particular, the scheme of a data attribute includes data type, value range, and formatting restrictions, but *not* knowledge that is specific for the outsourced data such as the value distribution of an outsourced attributes values.

  Example: The knowledge that the attribute *mark* has the value domain of {1,2,3,4,5,6} constitutes BKS-knowledge. An attacker that knows the value domain of *mark* but does not have background knowledge that is specific for the outsourced data such as the value distribution of *mark* constitutes a BKS-attacker.

- *Background knowledge on the data's content (BKD)*: If an attacker has background knowledge on the data's content we call her BKD-attacker. Background knowledge on the data's content can origin from many sources including common knowledge and other databases that are accessible by the attacker. Note that BKS knowledge can be inferred from BKD knowledge.

  Example: The knowledge that the values of an attribute *age* are normally distributed in a set of data constitutes BKD-knowledge. BKS-knowledge such as the value domain of *age* can be derived based on the distribution of *age*.

- *Background knowledge on the data and executed queries (BKQ)*: If an attacker has background knowledge on the data's content and on the executed queries we call her BKQ-attacker. In particular, BKQ-attackers can potentially derive confidential information from query access patterns.

  Example: The knowledge that two identical queries executed in a row implies that a specific record contains a specific attribute value constitutes BKQ knowledge.

In our model which assumes honest-but-curious attackers, attackers can be categorized according to what they are able to observe and their background knowledge. We show how it can be assessed whether a CPI provides protection against each attacker model in the next section.

3.4.2 Methodology to Assess CPIs

To determine whether it is suitable to apply a specific CPI in a given deployment scenario, the CPI has to be assessed with regard to the introduced taxonomy. It has to be determined whether the data is sufficiently protected by the CPI, whether it protects against the assumed attackers in the deployment scenario, and whether it provides the required query functionality. Most CPI publications clearly state the supported functionality and the achieved protection levels. However, it is not always stated against which attacker models the CPI provides the specified protection level. Instead, CPI publications often provide the proof that certain security properties are guaranteed. It depends on both the security property of a CPI *and* the attacker's capabilities whether a CPI can be considered secure. In the following, we first show the most common security properties of CPIs and then show which security properties are required to protect against each possible attacker model in our attacker taxonomy. This mapping allows to determine against which attacker a CPI can provide protection. We differentiate between the following security properties.

- *Content confidentiality*: The representation of the outsourced data at the CSP has a significant impact on whether an attacker with access to the outsourced data and background knowledge is able undermine the confidentiality of outsourced records. We differentiate between CPIs that store *order-preserving ciphertexts*, *distinguishable ciphertexts*, and *indistinguishable ciphertexts* at the CSP. Order-preserving ciphertexts hide plaintext attribute values but leak the order of the plaintext values, i.e., it holds that $enc(A) > enc(B)$ if $A > B$ for all values A and B. Distinguishable ciphertexts do not leak the order but can be distinguished from each other, i.e., it is possible to distinguish between ciphertexts that encrypt value A from ciphertexts that encrypt value B. Equal plaintext values are encrypted to equal ciphertext values. In particular, it is possible to determine which distinguishable ciphertexts contain the same plaintext value without decrypting the ciphertexts. Indistinguishable ciphertexts cannot be distinguished as encrypting a value A twice results in different ciphertexts that cannot be distinguished from a ciphertext that resulted from encrypting value B.

  Note that there exists an order in the protection guarantees of the ciphertext categories: indistinguishable ciphertexts provide all the protection guarantees of distinguishable ciphertexts. Distinguishable ciphertexts provide all the protection guarantees of order-preserving ciphertexts.

- *Access confidentiality* [CdVFP+13]: A CPI enforces access confidentiality if it guarantees that an attacker who observes a query cannot determine the records the query aimed for. Thus, for SELECT queries it has to be impossible for an attacker to determine which records are the results of the query. For INSERT, UPDATE, and DELETE queries an attacker must not be able to determine which records were modified or deleted.

- *Pattern confidentiality* [CdVFP+13]: A CPI enforces pattern confidentiality if it guarantees that it is impossible for an attacker to recognize query patterns. In particular, this implies that an attacker must not be able to determine that the same query has been executed multiple times in a row.

For CPI users without cryptographic expert knowledge it can be hard to specify the required security properties. It is typically easier for them to specify which attackers should be considered than to specify which low-level security properties are required. To facilitate the specification of deployment scenario requirements, we show which security properties are required to provide protection against specific attacker capabilities, i.e., we show how security properties can be mapped on attacker models of our taxonomy. This allows an easy assessment of CPIs with regard to which attacker models they provide protection against. Once the CPIs are assessed this way, deployment scenario requirements with regard to CPI security properties do not have to be specified but it suffices to specify the assumed attackers.

An overview of our methodology to determine the attacker models a CPI provides protection against is shown in Figure 3.3. A CPI that provides computational record protection against a given attacker model has to satisfy the security properties that cover the corresponding table cell.

At least **order-preserving ciphertexts** are required to enforce confidentiality against NBK-attackers with no background knowledge. As in our taxonomy order-preserving ciphertexts offer the least protection, the alternative would be to outsource plain-text values.

> *Example:* In Table 3.2b an attacker can observe that Bob has a better rating than Adam based on the order-preserving *rating* ciphertexts. However, without any background knowledge like the value domain or frequency distribution of the attribute *rating* an attacker is not able to reveal the plaintext rating of Adam and Bob.

At least **distinguishable ciphertexts** are required to enforce confidentiality against BKS-attackers with background knowledge on the data's scheme. Based on data scheme information like the value domain of an attribute, it is possible to reveal plaintext values from order-preserving ciphertexts as the following example shows.

> *Example:* In Table 3.2b an attacker who knows that {1,2,3,4,5,6} is the value range of *rating* (BKS knowledge) can determine plaintext ratings based on the order-preserving ciphertexts. For instance, Adam has to have rating 1 as his order-preserving ciphertext has the lowest value and the encrypted data contains five other distinct ciphertexts. Based on distinguishable ciphertexts this attack would no longer be possible. For instance, if only the scheme of *position* (it can take the values 1, 2, or 3) is known to the attacker and she has no background knowledge on the data's content, it is not possible to determine any plaintext *position* value in Table 3.2b.

The guarantee that only distinguishable ciphertexts are stored at the CSP is not sufficient to enforce computational record protection if range queries are executed and can be observed by a BKS-attacker. In this case, a CPI has to additionally guarantee access confidentiality, i.e., an attacker must not be able to determine which records are the results of an observed query.

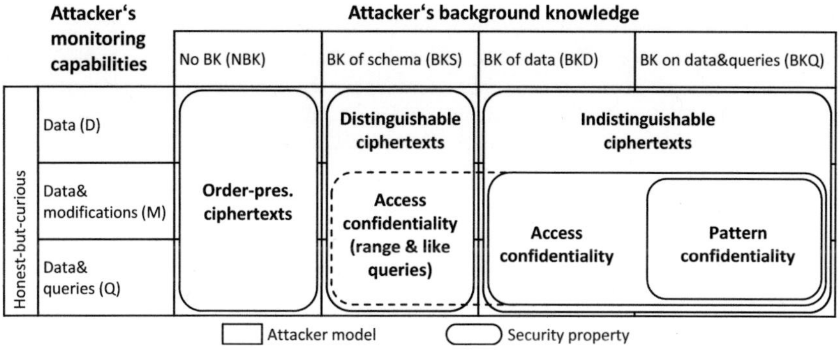

Figure 3.3: The *Attacker-Security Property Mapping* shows which security properties a CPI has to satisfy in order to protect against specific attackers. An approach that provides computational record protection against a given attacker model has to satisfy the security properties that cover the corresponding table cell [KJH15].

> *Example:* Consider that an attacker observes that records β, γ, δ, ϵ, and ϕ are returned when executing one range query and records α, β, γ, δ, and ϕ are returned when executing another range query on Table 3.2b. An attacker who has the BKS knowledge that the *position* value domain is $\{1,2,3\}$ can determine that the records β, γ, δ, and ϕ have the position 2 since it is the only possible explanation for the observed query results. Note that this attack works regardless of whether the *position* ciphertexts are distinguishable or indistinguishable. Thus, the only possibility to avoid this attack is to prevent the attacker from learning which records are returned, i.e., to provide access confidentiality.

The exemplary attack can also be performed by attackers that can observe modifications of the data. For instance, if records that contain attribute values within a certain value range are updated, the attacker can determine which records matched the update query by comparing the database before the update with the database after the update.

At least **indistinguishable ciphertexts** are required to enforce confidentiality against attackers with background knowledge on the data's content or the data's content and executed queries. Based on BKD knowledge like the frequency distribution of attribute values it is possible to derive plaintext values from distinguishable ciphertexts. Based on indistinguishable ciphertexts this attack is no longer possible as the ciphertext are not distinguishable from random values for the attacker and, thus, background knowledge such as the occurrence frequency cannot be applied.

> *Example:* In Table 3.2b an attacker who has background knowledge on the data's content and knows that the majority of records have 2 as a *position* value can use this knowledge to reveal *position* attribute values. As η is the most common distinguishable ciphertext it maps to 2. This

constitutes a so-called frequency attack. This attack is not possible based on indistinguishable ciphertexts. For instance, even if it is known to the attacker that 20000 is the most frequent *salary* in the data, she cannot apply this knowledge on the ciphertexts in Table 3.2b as it is not possible to tell which ciphertext encrypt the same plaintext value based on indistinguishable ciphertexts.

The guarantee that only indistinguishable ciphertexts are stored by the CSP is not sufficient in case an attacker can observe executed queries or modifications that are performed on the data. If an attacker knows which records match a query, she can distinguish the resulting records from the others and apply BKD knowledge again. To prevent the attacker from distinguishing the ciphertexts based on observed queries, access confidentiality is required.

If the attacker also has background knowledge on queries, a CPI also has to enforce pattern confidentiality to enforce data confidentiality. For instance, an attacker might know that if a query is executed twice in a row, a specific record has certain confidential attribute values. Thus, if the attacker observes that a query is executed twice in a row, data confidentiality is undermined. This attack can be avoided by CPIs that hide the query pattern from the CSP, i.e., that provide pattern confidentiality.

3.4.3 Survey of Existing CPIs

We apply the proposed methodology to categorize existing CPIs according to the taxonomy that we provided in Section 3.4.1. We introduce the resulting CPI catalog in this section. The taxonomy can be used both to categorize CPIs and to express deployment scenario requirements. Thus, the CPI catalog allows to determine which CPIs can be applied to satisfy given deployment scenario requirements.

The final CPI catalog is shown in Table 3.3. Note that some CPIs protect against multiple attacker models depending on the utilized functionality of the CPI. In the following we introduce various CPIs and explain their categorization in the CPI catalog.

Deterministic Indexes: In order to enable the CSP to evaluate equality selections on an attribute, deterministic substitutes of plaintext values can be outsourced instead of the plaintext values. For instance, keyed hash functions or deterministic encryption schemes that produce equal ciphertexts for equal plaintexts can be used to generate such deterministic substitutes. To search for a plaintext value X, it has to be replaced by its deterministic substitute within the query before the query is passed to the CSP.

Example: To execute the query `SELECT ... WHERE position=1` on Table 3.2b, the mediator has to generate the deterministic substitute θ of the attribute value "1" by using the secret encryption key. For instance, this can be achieved by applying a keyed hash function on "1". The mediator then substitutes the plaintext value in the original query with the deterministic substitute and passes the query `SELECT ... WHERE position=`θ to the CSP. The CSP can evaluate this query based on Table 3.2b and return the encrypted record with the ID α as a result.

CPI approach	Satisfiable deployment requirements															CPI Security Properties
	Functionality							Prot. goal	Attacker model							(Properties in brackets are only provided to a certain degree.)
	Queries				Modif.				Monit.			Knowl.				
	ES	RS	LS	AG	Insert	Update	Delete		Data	Mod.	Quer.	NBK	BKS	BKD	BKQ	
Deterministic Indexes	X				X	X	X	C	✓	✓	✓	✓	✓			Distinguishable ciphertexts
Deterministic Indexes (Flattened) [CDV+05]	X				X	X	X	P	✓			✓	✓	✓	✓	(Indistinguishable ciphertexts)
Bucketization [HILM02, HMCK12]		X			X	X	X	C	✓			✓	✓			Distinguishable ciphertexts
		X			X	X	X	C	✓	✓	✓	✓				
		X			X			C	✓	✓		✓	✓			
Bucketization (Flattened) [HILM02, HMCK12]		X			X	X	X	P	✓			✓	✓	✓	✓	(Indistinguishable ciphertexts)
Order-Preserving Encryption [AKSX04, BCLO09]	X	X			X	X	X	C	✓	✓	✓	✓				Order-preserving ciphertexts
Searchable Encryption [KPR12, SWP00, CM05, BBO07, CM05, KV08, LO05]	X	X	X		X	X	X	C	✓			✓	✓	✓	✓	Indistinguishable ciphertexts
	X	X	X		X	X	X	C	✓	✓	✓	✓				
	X				X	X	X	C	✓	✓	✓	✓	✓			
	X				X			C	✓	✓		✓	✓	✓		
Encrypted B-Trees [CDV+05, DDJ+03]	X	X	x[a]		X	X	X	C	✓			✓	✓	✓	✓	Indistinguishable ciphertexts
	X	X	x[a]		X	X	X	C	✓	✓	✓	✓				
	X	X	x[a]		X			C	✓	✓		✓	✓	✓		
Encrypted B-Trees (Shuffled) [CdVFP+13]	X	X	x[a]		X	X	X	P	✓	✓	✓	✓	✓	✓	✓	(+Access & pattern confidentiality)
Fragmentation [For11]	X	X	X	X	X	X	X	P	✓			✓	✓	✓	✓	(Indistinguishable ciphertexts)
Fragmentation (Non-colluding SPs) [ABG+05, HHK+10, GTF+11]	X	X	X	X	X	X	X	C	✓	✓	✓	✓	✓	✓		Indistinguishable ciphertexts Access confidentiality
Homomorphic Encryption [Pai99, OU98, RAD78, HIM04, MNT06]				X	X	X	X	C	✓	✓	✓	✓	✓	✓		Indistinguishable ciphertexts Access & pattern confidentiality
Oblivious RAM [GO96, SvDS+13]	X	X	x[a]		X	X	X	C	✓	✓	✓	✓	✓	✓	✓	Indistinguishable ciphertexts Access & pattern confidentiality
Oblivious RAM (Non-colluding SPs) [SS13]	X	X	x[a]		X	X	X	C	✓	✓	✓	✓	✓	✓	✓	
Private Information Retrieval [AMG07, KO97, CG97, SC07]	X	X	x[a]					C	✓	✓	✓	✓	✓	✓	✓	Indistinguishable ciphertexts Access & pattern confidentiality
Private Information Retrieval (Non-colluding SPs)[BS03, CKGS98, BIKR02]	X	X	x[a]					C	✓	✓	✓	✓	✓	✓	✓	

[a] Like selections are supported to a limited degree (e.g., prefix matching).

Legend

ES: Equality selection
RS: Range selection
LS: Like selection
AG: Aggregation

NBK: No background knowledge
BKS: Background knowledge of the data's schema
BKD: Background knowledge of the data's content
BKQ: Background knowledge of the data's content and queries

C: Computational record protection
P: Probabilistic record protection

Table 3.3: CPI catalog [KJH15].

Determining the plaintext value that maps to a deterministic substitute without knowing the secret key is infeasible. Deterministic substitutes are distinguishable and only equality selections are evaluated. Thus, based on Figure 3.3, deterministic index CPIs can protect against Q-BKS-attackers, i.e., attackers who have background knowledge on the data's scheme and can eavesdrop queries.

In order to also apply deterministic indexes against D-BKD-attackers, i.e., attackers with background knowledge on the data's content such as the frequency distribution of an attribute, *flattened hash indexes* were proposed [DDJ+03]. Flattened hash indexes map different plaintexts to the same deterministic substitute so that ideally each deterministic substitute occurs the same number of times. Thus, background knowledge on an attribute value's frequency in the data can no longer be applied. However, it is still possible to narrow down possible value candidates for attributes of specific records due to the fact that the same plaintexts still map to the same ciphertexts.

> *Example:* Even if each deterministic substitute for *position* would occur the same number of times in Table 3.2b for an attacker it would be possible to narrow down the possible positions of Bob if she knew that Adam has position "1". As Adam maps to θ, position "1" maps to θ. As Bob maps to η, it is impossible that Bob maps to position "1".

Thus, flattened hash indexes can be used to protect against B-BKD-attackers, but only provide probabilistic record protection as the possible attribute values of records can be narrowed down based on BKD knowledge.

Bucketization: By applying bucketization [HILM02, HMCK12], deterministic indexes can be extended so that the CSP can participate in the evaluation of range selections. Bucketization CPIs sort plaintext values into buckets which represent continuous value ranges. Each bucket is assigned an ID that does not reveal any information of the bucket range and the contained values. For each plaintext value, the according bucket ID is outsourced instead of the plaintext value. To execute a query that contains a range selection on an attribute, a query that selects each bucket that intersects with the queried range is sent to the CSP. By doing this, false-positive records may occur that have to be filtered out by the mediator.

> *Example:* Given the buckets $\{z = [1,3], d =]3,5], p =]5,7]\}$ for attribute A, a query `SELECT ... WHERE A<4` would be rewritten by the mediator to `SELECT ... WHERE A=z OR A=d`. As the CSP knows the bucket IDs of each record, it can evaluate the query. However, false-positive records with A=5 are also returned as bucket d contains values in the range of $[3,4]$ as well as in $]4,5]$.

Since the bucket IDs are distinguishable to the CSP and range queries are evaluated, bucketization can protect against D-BKS-attackers (see Figure 3.3). Bucketization also protects against Q-NBK-attackers. Based on eavesdropped queries, the order of bucket IDs can be determined. Thus, the distinguishable ciphertexts can be transformed to order-preserving ciphertexts by a Q-attacker who can use knowledge on the data's scheme (BKS knowledge) to reveal plaintext values (see Section 3.4.2).

Like in the case of deterministic indexes, buckets can be flattened to avoid the application of background knowledge on the attribute frequency. For the same reasons as with deterministic indexes, flattened bucketization CPIs can only provide probabilistic record protection against D-BKQ-attackers.

Order-Preserving Encryption: To avoid the transmission of false-positive records, order-preserving encryption (OPE) schemes [AKSX04, BCLO09] can be used to generate and outsource order-preserving ciphertexts instead of the plaintext values. OPE schemes maintain the order of plaintext values when they generate ciphertexts. This allows a CSP to evaluate < and > operators on the ciphertexts without knowing the encryption key. To execute a range query based on order-preserving ciphertexts, the mediator has to replace the plaintext values in the query with the according order-preserving ciphertexts.

> *Example:* To evaluate the query `SELECT ... WHERE rating>4` on Table 3.2b, the mediator has to replace "4" with the according order-preserving ciphertext "742". The rewritten query `SELECT ... WHERE rating>742` can be evaluated by the CSP based on Table 3.2b.

Since order-preserving ciphertexts are outsourced to the CSP, OPE schemes can protect against Q-NBK-attackers, i.e., attacker without any background knowledge (see Figure 3.3).

Searchable Encryption: Searchable encryption (SE) schemes [KPR12, SWP00, CM05, BBO07, KV08, LO05] can be used to encrypt attribute values and outsource indistinguishable ciphertexts. SE schemes allow the mediator to generate predicate tokens based on the encryption key. These predicate tokens can be used by the CSPs to check whether the plaintext value of a ciphertext matches a specific predicate without decrypting the ciphertext. Predicates can encode equality, range and like selections. Furthermore, a predicate token does not reveal the predicate that is checked to CSPs. Thus, SE schemes enable CSPs to evaluate query conditions without enabling the CSPs to decrypt the data or revealing the predicates they check.

> *Example:* To search for records that match the predicate *salary*=20000 in Table 3.2b a mediator would generate a predicate token X for the predicate and send it to the CSP. The CSP can use this token to check which of the *salary* ciphertexts match the predicate without actually knowing for which predicate it checks the records. X matches for ι, κ, and μ. Thus the CSP returns records α,β, and δ.

Since the ciphertexts of SE schemes are indistinguishable, SE schemes can protect against D-BKQ-attackers who have access to the data (see Figure 3.3). Based on eavesdropped queries or data modifications, an attacker can monitor which records and attribute values match a query and therefore distinguish them. Background knowledge on the data's content can be applied to distinguishable ciphertexts to reveal attribute values. SE schemes do not provide access confidentiality [IKK12] and, therefore, cannot prevent attackers that eavesdrop queries from distinguishing ciphertexts. Thus, if an attacker can eavesdrop queries, SE schemes only provide protection if the attacker has no background knowledge on the data's content (Q-BKS-attacker). Additionally, if range or like queries are evaluated, SE schemes can only provide protection against Q-NBK-attackers (see Figure 3.3).

Encrypted B-Trees: B-Trees are utilized by traditional database systems to index data and allow an efficient execution of equality and range selections as well as prefix matching selections. Indistinguishably encrypting each of the B-Trees nodes was proposed to enforce data confidentiality [CDV+05, DDJ+03]. Since the encrypted B-Tree contains only indistinguishable ciphertexts it has to be maintained by the mediator rather than the CSP. Furthermore, the mediator has to participate actively when traversing the B-Tree to find the records that match the query condition. To find the records that match a condition, the mediator retrieves the root node of the B-Tree, decrypts it and descends into the tree as in the unencrypted case. All nodes on the path from the root of the B-Tree to the leaf which contains the target records are retrieved and decrypted. Thus, log(n) rounds of communication are required to retrieve the records that match the selection conditions.

Since the outsourced ciphertexts are indistinguishable, encrypted B-Trees can protect data confidentiality against D-BKQ-attackers (see Figure 3.3). An attacker that eavesdrops queries or modifications can distinguish ciphertexts and infer the order of the ciphertexts due to the B-Tree's ordered structure [PZM13]. Thus, as shown in Figure 3.3 B-Trees can only protect against attackers without any background knowledge if the attackers can monitor queries or modifications (Q-NBK-attackers).

In order to enhance encrypted B-Trees to also protect against attackers with background knowledge who are able to eavesdrop queries it was proposed to enforce access and pattern confidentiality by shuffling the B-Tree after each query [CdVFP+13]. For each node that is retrieved from the B-Tree, *e* nodes from the same level of the B-Tree are retrieved. Each of the additionally retrieved *e* nodes is a so-called cover search. Once the leaf nodes are retrieved, the *e* nodes of each level are re-encrypted and shuffled by the mediator and written back to the B-Tree. Thus, if the same query is executed again, the access pattern on the B-Tree looks different from the previous one (pattern confidentiality) and the CSP cannot be sure which records are actually retrieved (access confidentiality). Since only *e* nodes are shuffled in each level of the B-Tree, the CSPs are still able to distinguish queries with a certain probability. Thus, we argue that shuffled B-Trees protect against strong Q-BKQ-attackers but only provide probabilistic record protection.

Fragmentation: To protect the confidentiality of attribute value combinations rather than values of a single attribute, fragmentation approaches [For11, ABG+05, HHK+10, GTF+11] that split records in multiple, unlinkable fragments were proposed. As the attribute values are not encrypted, CSPs can evaluate equality, range and like selections as well as perform aggregations. However, queries with multiple conditions on different attributes that are contained in different fragments cannot be completely evaluated by the CSP.

Example: The records in Table 3.2b are fragmented to protect the attribute combination *pseudonym* and *age*. An attacker is not able to link the fragments, as the *ID* attribute values are indistinguishably encrypted in both fragments. To evaluate a query `SELECT ... WHERE (pseudonym=Adam OR pseudonym=Bob) AND (age=23 OR age=30)`, the CSP either considers only the condition on *pseudonym* or evaluates each query condition on the according fragment and lets the mediator perform a join. The first option leads to the transmission of the records α and β. Once the records are decrypted by the mediator, the additional condition `age=23 OR age=30` is evaluated and the β record is discarded as a false-positive record. The second option leads to the transmission of the record IDs α, β that match the condition on *pseudonym* and π, τ, υ, ϕ that match the condition on *age*. These record IDs are decrypted by the mediator and joined, so that only the record ID "1" remains. The according record is then retrieved from the CSP and decrypted in a second step. Both options can lead to significant overheads that originate from joining and transmitting false-positive records as we show in Section 3.5.6.

The data fragments can be stored at a single CSP or multiple, non-colluding CSPs. In case the fragments are stored at a single CSP [For11], attackers that are able to eavesdrop modifications can link the fragments. For instance, if a new record is inserted, the attacker can infer that the newly added partial records in each fragment belong together. Thus, single CSP fragmentation approaches only protect the confidentiality of attribute combinations against D-BKQ-attackers that cannot eavesdrop modifications. Furthermore, even if the attributes that are not part of a fragment can be considered as indistinguishable ciphertexts in the sense that it is not possible to distinguish them and apply background knowledge, the attacker still learns more from the outsourced fragments than from random attribute values. In particular, attackers can narrow down possible attribute value candidates for a record. Thus, single CSP fragmentation approaches only provide probabilistic record protection.

Example: An attacker who has access to the data shown in Table 3.2b can determine that Adam is either 23, 25, or 30 years old even if she is not able to link the fragments and determine Adam's exact age. Therefore, she is able to narrow down the possible *age* attribute values for each record.

In case the fragments are stored at multiple, non-colluding CSPs [ABG+05, HHK+10, GTF+11], a CSP only has access to a single fragment. Thus, only the changes that are performed on a single fragment can be observed by eavesdropping modifications and linking record fragments by correlating the changes on multiple fragments is not possible. Furthermore, the attribute values of attributes that are not contained in a fragment can be considered truly indistinguishable for a CSP as it cannot access any values of the attribute. Thus, multi CSP fragmentation approaches can provide computational record protection against Q-BKD-attackers.

Homomorphic Encryption: Applying homomorphic encryption schemes [Pai99, OU98, RAD78] enables CSPs to aggregate indistinguishable ciphertexts without knowing the encryption key [HIM04, MNT06]. The execution of specific operations in the ciphertext domain of homomorphic encryption schemes has the effect that specific operations such as addition or multiplication are executed in the plaintext domain. Thus, it is possible to execute calculations on ciphertexts without decrypting them. Based on this property, the CSP can perform aggregations on ciphertexts as the following example shows.

> *Example:* A mediator encrypts the values "2" twice and outsources the resulting ciphertexts ω and γ to the CSP. The CSP is not able to determine that ω and γ are ciphertexts of the same plaintext. Furthermore, it can aggregate them to the new ciphertext ι without knowing the encryption key. The mediator knows the encryption key, decrypts ι and gets the plaintext "4" as result.

The ciphertexts of homomorphic encryption schemes are indistinguishable. Furthermore, access confidentiality is naturally given as aggregation queries do not select specific records but are executed on all records or on a set of records that is selected by other CPIs. Thus, homomorphic encryption can protect against Q-BKD-attackers with background knowledge on the data's content as well as the ability to monitor queries (see Figure 3.3).

Oblivious RAM: Oblivious RAM (ORAM) approaches [GO96, SvDS$^+$13, SS13] outsource indistinguishable ciphertexts and ensure that queries on these ciphertexts are indistinguishable. ORAM approaches achieve this by shuffling the outsourced data after each query or modification in such a way that a computationally bounded attacker does not have a better chance of determining whether two queries were different or similar than guessing. As queries are indistinguishable, Oblivious RAM approaches enforce access and pattern confidentiality. Therefore, ORAM CPIs can provide computation record protection against strong Q-BKQ-attackers that can eavesdrop queries and have background knowledge on the data's content and queries (see Figure 3.3).

Private Information Retrieval: Private Information Retrieval (PIR) approaches [CKGS98, OG10] allow to execute queries on outsourced data that are indistinguishable to the CSP. Therefore, the enforce pattern and access confidentiality like ORAM and provide computational record protection against Q-BKQ-attackers. However, they can only be applied to retrieve data and normally do not allow to modify the outsourced data. Unlike ORAM they allow to access a block of data in a single communication round. PIR approaches can achieve this at the expense of an increased computation overhead at the CSP (computational PIR) [AMG07, KO97, CG97, SC07] or by assuming non-colluding CSPs (information-theoretic PIR) [BS03, BIKR02, CKGS98]. To build indexes based on PIR, additional indexing structures that are accessed via PIR are required. For instance, encrypted B-Trees can be combined with PIR. However, this would imply multiple communication rounds and cancel out the benefit that PIR approaches have over ORAM approaches.

3.4.4 Efficiency

In this section we show that aside from the differences regarding the functionality, protection levels and addressed attacker models, CPIs differ significantly with regard to efficiency. We show this by analytically assessing the CPIs. We investigate the *transmission overhead*, i.e., the amount of data that has to be transmitted from the CSP to the mediator when executing a query. We also analyze the *number of communication rounds* that have to be sequentially performed between the CSP and the mediator to evaluate a query. The number of communication rounds has a substantial impact on query latency as they result in stacked network latency. Furthermore, we investigate the *number of touched records* that have to be touched by a CSP to evaluate a query and that result in I/O overhead. Finally, we provide a rough estimation of the computational overhead that CPIs induce at the mediator and the CSPs. To assess computational overhead, we assign the level *none* if no or almost no calculations have to be performed, *low* if only few lightweight cryptographic operations such as hash functions have to be executed, *moderate* if only a few cryptographic operations based on symmetric encryption schemes have to be performed, and *high* if more resource-consuming asymmetric encryption schemes or a large amount of lightweight operations are required. Building detailed performance models for CPIs is beyond the scope of this thesis. However, we provide exemplary empirical measurements in Section 3.5.6 that provide an impression on the practical performance impact of selected CPIs.

The results of our analysis are shown in Table 3.4. In the following we highlight the most important findings that underline the substantial performance differences that exist between CPIs.

Overall, deterministic indexes and fragmentation induce the least overhead. Fragmentation approaches require the transmission of false-positive records if the conditions of a query cannot be evaluated based on attributes that are contained in a single fragment. This is more of a problem for single CSP fragmentation approaches, as they do no allow to store an attribute in multiple fragments to avoid linking attacks. We investigate the impact of the transmission overhead induced by fragmentation approaches further in Section 3.5.6.

Flattened deterministic indexes and bucketization CPIs also induce false-positive result records and therefore increase the transmission overhead as well as the number of records that the CSPs must touch. The number of false-positive records highly depends on the distribution of the outsourced attribute values. For instance, consider a scenario in which n/2 records have attribute value a, n/4 records have attribute value b, and n/4 records have attribute value c. For a flattened deterministic index where b and c map to one deterministic substitute and a maps to another, the number of false-positives when querying for records that contain attribute value b would be $n/4$.

Encrypted B-Trees require log(n) consecutive communication rounds between the mediator and the CSP to find the records that match a query condition. In particular, this leads to a stacking of the network latency between the mediator and the CSP for each round of communication.

	Transmission Overhead	Comm. Rounds	Touched Records	Computation CSP	Computation Mediator
Det. Indexes	O(k)	1	O(k)	none	low
Det. Indexes (Flattened)	O(k+e)	1	O(k+e)	none	low
Bucketization	O(k+e)	1	O(k+e)	none	low
Bucketization (Flattened)	O(k+e)	1	O(k+e)	none	low
Order-Preserving Encryption	O(k)	1	O(k)	none	low
Searchable Encryption	O(k)	1	O(n)	moderate	moderate
Encrypted B-Trees	O(log(n)+k)	log(n)	O(log(n)+k)	none	moderate
Encrypted B-Trees (Shuffled)	O(2·e·log(n)+k)	log(n)+1	O(2·e·log(n)+k)	none	moderate
Fragmentation	O(k+e)	1	O(k+e)	none	none
Fragmentation (non-colluding SPs)	O(k)	1	O(k)	none	none
Homomorphic Encryption	O(1)	1	O(k)	high	high
Oblivious RAM	Depends on the specific approach				
Example [SvDS+13]	$O(\log(n)^2 + \log(n)\cdot k)$	log(n)+1	$O(\log(n)^2+\log(n)\cdot k)$	none	moderate
Example [SS13] (non-colluding SPs)	SP→Client: O(log(n)+k) SP→SP: $O(\log(n)^2+\log(n)\cdot k)$	log(n)+1	$O(\log(n)^2+\log(n)\cdot k)$	moderate	moderate
Private Information Retrieval	Depends on the specific approach				
Example [BIKR02, OG10] (non-colluding SPs)	Hash Index: $O(n^{\frac{1}{3}}) + k$ B-Tree: $O(\log(n)\cdot n^{\frac{1}{3}} + k)$	1 log(n)	$O(n^{\frac{1}{3}}) + k$ $O(\log(n)\cdot n^{\frac{1}{3}} + k)$	low	low

n: Number of outsourced attribute values
k: Number of records matching a query
e: Number of (unnecessarily transmitted) false-positive records

Table 3.4: Performance categorization of existing approaches[KJH15].

Searchable encryption requires the CSP to apply the predicate token on all outsourced ciphertexts that are contained in an index to evaluate a query. Thus, compared to the sub-linear search times of other CPIs like B-Trees or deterministic indexes, searchable encryption require a linear search time with regard to the number or outsourced attribute values.

Homomorphic encryption induces high computational overheads compared to other CPIs. In this thesis, we only use partially homomorphic encryption. The overhead for fully homomorphic encryption is much higher. In Section 3.5.6 we provide measurements that capture the overheads induced by homomorphic encryption in a practical scenario.

There exists a variety of ORAM and PIR approaches that each make their own trade-offs between transmission and computational overhead as well as the required communication rounds. We exemplarily adapted ORAM/PIR protocols that are considered to be among the most efficient protocols for the database outsourcing scenario and analyzed their efficiency [Deg15]. They can be considered very expensive compared to other CPIs with regard to transmission overhead, the required rounds of communication and the number of records that have to be touched to evaluate a query.

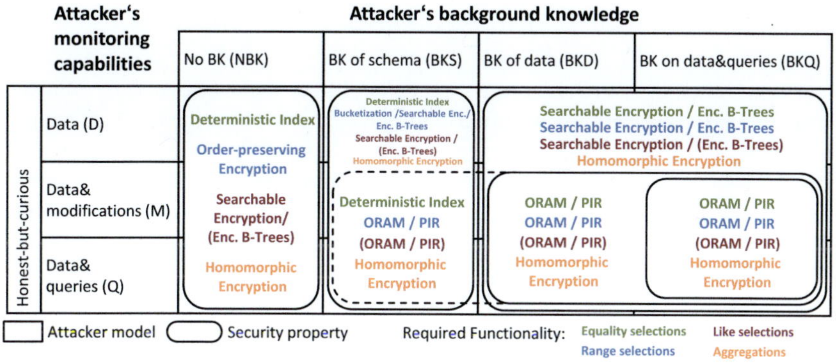

Figure 3.4: Efficiency-optimal CPIs that offer the required functionality and the security properties that are required to provide computational record protection against each attacker model (fragmentation excluded) [KJH15].

3.4.5 Conclusions

We categorized existing CPIs according to the introduced taxonomy based on the protection levels a CPI achieves against various attacker models and their ability to execute specific kinds of queries efficiently. Based on this categorization we can make recommendations on which CPIs should be used to enforce the confidentiality of a given attribute and allow an efficient evaluation of query conditions on the attribute at the same time. Our recommendations are summarized in Figure 3.4. For each type of query functionality the most efficient CPIs are listed that can be used to provide computational record protection against the given attacker model. In some cases multiple CPIs are listed as the most efficient CPI choice depends on the deployment scenario. For instance, if network latency is negligible, B-Trees can outperform searchable encryption as multiple communication rounds do not have a substantial impact on efficiency. Fragmentation approaches were omitted in Figure 3.4 as fragmentation alone is typically insufficient to provide computational records protection due to the fact that too few CSPs are available. However, we show in Section 3.5 how non-colluding CSPs can be leveraged as much as possible to avoid efficiency overheads.

Based on the study we can only give recommendations on which CPIs should be used to enforce the confidentiality of a *single* attribute. In real deployment scenarios, databases contain a multitude of attributes. In many scenarios, different confidentiality requirements exist for different attributes. Furthermore, in some cases not the attribute values are considered confidential, but the combination of attributes. In this setting, choosing an efficiency-optimal set of CPIs to satisfy the deployment scenario requirements is much more complex than in the case with just one attribute. We address this advanced problem setting in Section 3.5.

3.5 Confidential Database-as-a-Service: Securus

Our conducted CPI study shows which CPIs protect the confidentiality of *single* attributes against specific attackers and allow an efficient evaluation of specific query conditions at the same time. However, in real deployment scenarios a database contains multiple attributes and it can be required to protect attribute combinations rather than single attributes. For each attribute in the database it can vary whether the confidentiality of the attribute has to be enforced. Furthermore, the background knowledge of the assumed attacker can vary for different attributes of a database. For instance, it could be okay to store attribute A in plaintext while only indistinguishable ciphertexts may be outsourced for a confidential attribute B due to existing background knowledge of the assumed attacker. Also CPIs do not have to be applied for all attributes as real query workloads often do not require to be able to evaluate every possible query condition on the data. For instance, no CPIs have to be applied on attribute B if its values are irrelevant for the conditions of the queries that will be executed. In particular, these observations lead to the following questions that have to be answered in real deployment scenarios:

- Which combination of CPIs is suitable to protect data confidentiality in a given scenario?
- Which suitable CPI combination for a scenario is optimal with regard to efficiency?

Designing an approach that is tailored to a deployment scenario is possible but requires cryptographic expert knowledge. If the approach should be applied to another deployment scenario or the deployment scenario requirements change, the approach has to be re-designed and re-implemented. To build a tunable approach that does not require to be re-designed and re-implemented by a cryptography expert for varying deployment scenario requirements, it is necessary to automatically derive a suitable set of CPIs from higher level policies that can be specified by a risk manager without deep cryptographic knowledge.

In this section we present *Securus* (**S**ecure and **E**fficient **C**loud **U**tilization **R**elying **U**pon **S**chemes), a tunable approach that preserves data confidentiality when outsourcing databases to CSPs. Securus addresses the research questions stated in Section 3.2.1: *How can an efficiency-optimized database outsourcing solution that satisfies confidentiality requirements be derived for a given query workload?*

Securus automatically generates a database outsourcing mediator that enforces high-level policies which describe data confidentiality and query requirements. To satisfy these policy requirements, Securus determines a suitable combination of CPIs and implements them in a software-component called mediator that can be used to seamlessly outsource and query data. This software-component applies the determined CPIs transparently when outsourcing data and executing queries on the outsourced data. Thus, risk managers can specify the confidentiality and query requirements without expert knowledge on CPIs. Furthermore, based on the generated mediator, client applications can outsource and query the data transparently without having to consider to apply CPIs. Securus can be tuned to satisfy different deployment scenario requirements as the automatic mediator generation avoids overheads for re-designing and re-implementing the outsourcing solution.

This section is organized as follows. We introduce the architecture of the Securus approach in Section 3.5.1. The following three chapters apply the methodology we proposed to build tunable approaches. We already identified the confidentiality-efficiency trade-off in the DaaS problem setting. In Section 3.5.2 we introduce Securus-Latin, a high-level language to specify deployment scenario requirements, including confidentiality and query requirements. We already identified and assessed security mechanisms that can be used to satisfy the requirements in Section 3.4. Based on these findings, we show how specified deployment scenario requirements can be satisfied via CPIs in Section 3.5.3. In Section 3.5.4 we show how a set of CPIs that satisfies the specified deployment scenario requirements can be automatically determined. In Section 3.5.5 we show how conflicting requirements can be detected and propose how such conflicts can be resolved. We highlight the tunability potentials of Securus in Section 3.5.6 and compare it to other existing database outsourcing approaches. We discuss the Securus approach in Section 3.5.7 and evaluate its performance. Finally, we conclude the section on confidential database outsourcing in Section 3.5.8.

3.5.1 Architecture

An overview of how database outsourcing is performed via Securus is shown in Figure 3.5. The database outsourcing process is comprised of three steps[4]:

1. The database architect and the risk manager specify the confidentiality requirements and define which queries should be efficiently executable in a so-called *policy profile* (PP).

2. Securus automatically determines a set of CPIs that satisfy the specified deployment scenario requirements that are specified in the PP and implements them in the mediator software-component.

3. The client application can use the mediator to transparently outsource the data to CSPs and query data, i.e., the user can send plaintext data and queries to the mediator and gets plaintext results in return. The mediator enforces the deployment scenario requirements that are specified in the policy profile.

Attacker model: We assume the CSPs to be *honest-but-curious*, i.e., they strive to gain confidential information from the outsourced data but do not manipulate any data or query results. To protect the outsourced data against malicious CSPs that manipulate data, additional integrity preserving techniques [MNT06, WYX13] can be integrated in Securus in future work. Furthermore, in case multiple CSPs are available we assume them to be *non-colluding*, i.e., the CSPs are assumed to be honest-but-curious on their own but do not collude with each other. Note that

[4] Note that step 1 and 2 only have to be performed once for each set of data.

Securus can also be used if only a single CSP is available. Furthermore, we assume that the attacker can observe queries but does not have background knowledge on the observed queries. We argue that the ORAM/PIR approaches, which are necessary to protect against attackers with background knowledge on queries, are not yet efficient enough to be applied for database outsourcing [Deg15]. Thus, we leave including ORAM/PIR approaches in Securus to protect against attackers with background knowledge on the queries for future work. Furthermore, we show in Appendix A.5 how Securus can leverage additional CPIs in certain situations if it can be assumed that the attacker cannot observe queries.

In Section 3.5.2, we show how PPs can be specified and show they can be satisfied based on CPIs in Section 3.5.3. In Section 3.5.4, we explain how the PPs can be transformed automatically into a mediator. User-specified policies may contradict each other which can result in unsolvable scenarios. We show in Section 3.5.5 how such conflicts can be isolated and resolved.

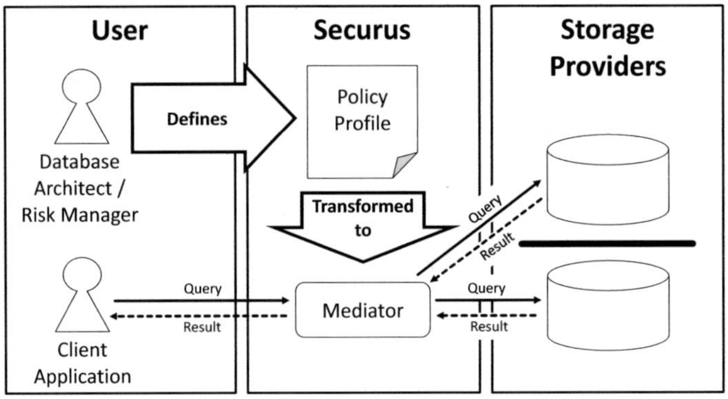

Figure 3.5: Architecture of Securus [KJ14].

3.5.2 Policy Profile Specification

Policy Profiles can be specified in the domain-specific language *Securus-Latin*. An exemplary PP is shown in Figure 3.6. A PP specifies the data structure, i.e., relational data tables that contain the attributes and their data types. Furthermore, a PP contains query requirements (*query policies*), confidentiality requirements (*confidentiality constraints*), and assertions on the assumed attacker (*inference constraints*).

Query Policies (QPs): a QP expresses that a certain category of queries has to be efficiently executable. Besides the *equality, range* and *like selections* that were introduced in Section 3.4, Securus supports other query constructs of the Structured Query Language (SQL) such as "GROUP BY" or "SORTED BY" components. Securus-Latin allows to specify the future query workload in SQL-like syntax as shown in Figure 3.6. The following example provides an impression on how to specify query policies to express which queries should be efficiently executable.

```
PolicyProfile EmployeeDB{
        Tables{
                Employees{
                        empID           :INTEGER
                        name            :STRING(10)
                        salary          :INTEGER
                        phoneNumber     :INTEGER
                }
        }
        QueryPolicies{
                SELECT * FROM Employees
                WHERE
                name = ? AND empID = ?
                SORTED BY ID DESCENDING;

                SELECT SUM(salary) FROM Employees
                WHERE
                empID < ?;
        }
        ConfidentialityConstraints{
                [name,salary]
                [name,empID]
        ]
        InferenceConstraints{
                DIC [empID]
                OIC [phoneNumber]
        }
}
```

Figure 3.6: Exemplary policy profile.

Example: The QP `SELECT * FROM Employees WHERE name = ? AND empID = ? SORTED BY ID DESCENDING` expresses that queries like `SELECT * FROM Employees WHERE name = 'Miller' AND empID = 123 SORTED BY ID DESCENDING` and `SELECT * FROM Employees WHERE name = 'Doe' AND empID = 512 SORTED BY ID DESCENDING` have to be efficiently executable by the mediator.

Confidentiality Constraints (CCs): A CC constitutes a subset of attributes and states that the association between values of the given attributes is sensitive. Thus, based on the outsourced ciphertexts, it must not be possible for any CSP to reveal and link plaintext values of all attributes that are contained in the CC at once. A CC that just contains a single attribute states that the values of the attribute must not be revealable by any CSP. The following example provides an impression on how to specify CCs to express confidentiality requirements. In Section 3.5.7, we show how CCs can be used to express confidentiality requirements in more detail.

Example: In many cases, not the attribute values themselves are considered confidential but the combination of them. For instance, it may be that neither the names of a company's employees nor the salaries the company pays are considered confidential. However, the combination of those two attributes, i.e., the information who earns how much money might be considered sensitive. A CC enforces that no CSP is able to link values of all the attributes contained in the CC. For instance, the CC `[name, salary]` enforces that no CSP can reveal both the *name* value and the *salary* value of a record.

The concept of CCs is versatile in the sense that it also allows to specify that single attributes have to be confidential by specifying singleton CCs. For instance, `[salary]` enforces that the attribute *salary* may not be viewed by any CSP.

Inference Constraints (ICs): CPIs can leak different kinds of information about the plaintext values of an attribute such as the order or the fact that the attribute values of two records are the same. Depending on the attacker's ability to apply background knowledge on this leaked information, it can be possible for her to reveal plaintext values (see Figure 3.3 in Section 3.4.2).

An IC makes an assertion on the ability of the attacker to apply background knowledge. We distinguish between *Order Inference Constraints* (OICs) and *Distinguishability Inference Constraints* (DICs). OICs make the assertion that if the order of given attribute values leak to the attacker, she is not able to reveal plaintext values of the according attribute. In particular, this is true if the attacker has no background knowledge on the scheme of an attribute (see Figure 3.3). For instance, the OIC `[A]` expresses that an attacker cannot reveal any plaintext values of *A* based on order-preserving ciphertexts of the attribute *A*. DICs express that the attacker can gain access to distinguishable ciphertexts of a given attribute without being able to reveal plaintext values of this attribute. In particular, this is true for attackers who do not possess background knowledge on the data (see Figure 3.3). The following example provides an impression on how to determine whether an inference constraint can be specified.

> *Example:* The DIC `[empID]` expresses, that no CSP can gain useful information from distinguishable ciphertexts of the attribute *empID*. In particular, this assumption can be made if the risk manager is sure that the assumed attacker lacks the necessary background knowledge on *empID* that can be applied on distinguishable ciphertexts to reveal plaintext values of *empID* (e.g., the frequency distribution of *empID*'s attribute values or plaintext *empID* values of certain records). Furthermore, it can be made independent from the attacker's background knowledge if the *empID*'s values are guaranteed to be unique as it is the case for user identifiers, for instance. If the values of an attribute are guaranteed to be unique, background knowledge such as the frequency distribution cannot be applied.

Securus generates a mediator that can be used to enforce the specified PP. In the following we will first show how a PP can be enforced based on CPIs in Section 3.5.3. Afterward, we show how Securus automatically generates mediators to enforce PPs in Section 3.5.4.

3.5.3 Policy Profile Satisfaction

To show how a policy profile can be satisfied by a mediator, we first show how an exemplary policy profile can be satisfied by applying CPIs. We then use the example to derive generic conditions that a mediator has to adhere to in order to satisfy a policy profile.

> *Example:* For the example policy profile that is shown in Figure 3.6 an exemplary CPI combination and the resulting representation of the outsourced data is shown in Figure 3.7. The data table *Employees* was outsourced to two non-colluding CSPs.
>
> CSP1 stores the attribute *name* in plaintext, indistinguishable AES ciphertexts of the attributes *age* and *salary* (*enc_empID* and *enc_salary* respectively), and keyed hashes of the attribute *empID* (*hash_empID*). Queries that are modeled by the query policy QP_1 can be efficiently executed by querying for the hashed value of *empID* instead of the plaintext value and decrypting the attributes *enc_empID* and *enc_salary* of the returned records at the mediator. Thus, the query policy QP_1 is satisfied. At the same time, none of the CCs `[name, empID]` and `[name, salary]` is violated as CSP1 only has access to *name* and neither to *salary* nor to *empID*. CSP1 cannot use *hash_empID* to break the CC `[name, empID]`

as the specified DIC [empID] guarantees that the distinguishable ciphertexts of *hash_empID* cannot be used by the attacker to reveal plaintext values.

CSP2 stores only the two attributes *empID* and *salary* in plaintext and can efficiently evaluate queries that are modeled by QP_2. Thus, the query policy QP_2 is satisfied. As CSP2 stores no attribute values for *name*, neither the CC [name,empID] nor the CC [name,salary] is violated at CSP2.

As both QP_1 and QP_2 are satisfied and no CCs are violated, the CPIs that are applied in Figure 3.7 satisfy the specified policy profile shown in Figure 3.6. Furthermore, the solution in Figure 3.7 leveraged the two non-colluding CSPs to avoid unnecessary encryption overhead and only applied CPIs that are strictly necessary to satisfy the policy profile.

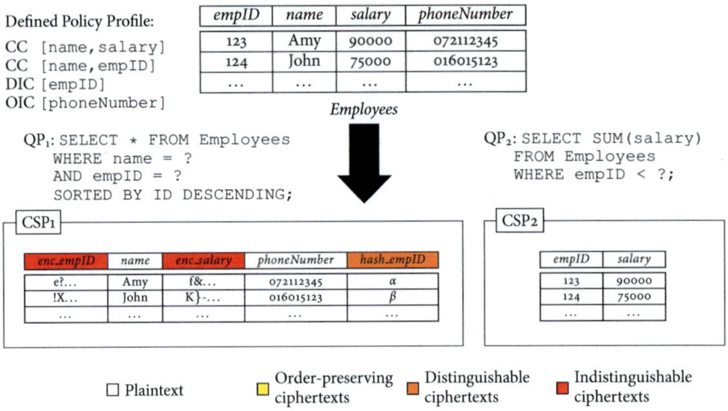

Figure 3.7: Exemplary policy profile enforcement.

The example in Figure 3.7 illustrates how a policy profile can be satisfied by applying CPIs. Whether a policy profile is satisfied by a set of CPIs depends on the ciphertext representations that are stored by each CSP. We denote the ciphertext representation at each CSP as attribute allocation.

Definition - Attribute Allocation: An attribute allocation states for each CSP which representation of each attribute may be stored.

An attribute allocation satisfies a policy profile if it satisfies the following conditions:

- **Condition 1:** For each QP at least one CSP has to store the relevant attributes in a representation that allows the CSP to execute efficiently the queries which are modeled by the QP. An overview of attribute representations that are required to evaluate different query components is shown in Table 3.5. The ✓ symbol indicates that there exists a CPI based on which the CSP can evaluate the according query component and that implies the according attribute representation.

		Attribute representation from a CSP's perspective			
Type	Query component	Plaintext	Order-preserving ciphertext	Distinguishable ciphertext	Indistinguishable ciphertext
1	Attr1 = ?	✓	✓ e.g., OPE [BCLO09]	✓ e.g., keyed hashes	
2	Attr1 < ?	✓	✓ e.g., OPE [BCLO09]		
3	Attr1 (NOT) LIKE ?	✓	✓ e.g., SE schemes [SWP00]		
4	Attr1 IN ?	✓	✓ e.g., OPE [BCLO09]	✓ e.g., keyed hashes	
5	SUM(Attr1)	✓	✓ e.g., Paillier [Pai99]	✓ e.g., Paillier [Pai99]	✓ e.g., Paillier [Pai99]
6	SUM(Attr1 · Attr2)	✓			
7	COUNT (DISTINCT Attr1)	✓	✓ e.g., OPE [BCLO09]	✓ e.g., keyed hashes	
8	GROUP BY Attr1	✓	✓ e.g., OPE [BCLO09]	✓ e.g., keyed hashes	
9	SORTED BY Attr1	✓	✓ e.g., OPE [BCLO09]		
10	SUBSTRING(Attr1)	✓			
11	HAVING SUM(Attr1) < ?	✓			
...	...	...	...	...	...

Table 3.5: Attribute ciphertext representations required to efficiently evaluate various query components.

Notice that the CPI mapping on query components and attribute representations that is shown in Table 3.5 can be applied to protect against attackers who can observe queries. For instance, searchable encryption (SE) schemes produce indistinguishable ciphertexts. However, when using SE schemes to evaluate queries of type 2, it has to be assumed that by observing queries an attacker is able to determine the order of ciphertexts (cf. Section 3.4). We show in Appendix A.5 how making the assumption that attackers are not able to observe queries changes the attribute representation of certain CPIs. That can be leveraged by Securus to satisfy policy profiles more efficiently. Furthermore, note that Securus currently does not consider ORAM/PIR CPIs as they exhibit poor performance even when compared to copying the entire database [Deg15]. They can be included in Securus once it is foreseeable that integrating them can have a benefit on query execution performance.

- **Condition 2:** All CCs have to be satisfied. A CC is considered satisfied if no CSP is able to view values for all attributes that are contained in the CC. The following conditions have to be satisfied to prevent a CSP j from viewing values of attribute i.

  - **Condition 2.1:** Plaintext values of attribute i must not be stored by CSP j.
  - **Condition 2.2:** If no DIC is specified that contains the attribute i, distinguishable ciphertexts of attribute i must not be stored by CSP j. If a DIC is specified for attribute i, storing distinguishable ciphertexts of this attribute at a CSP does not give the CSP the ability to view values for attribute i. This is due to the fact that the DIC guarantees that the CSP cannot reveal plaintext values based on distinguishable ciphertexts of attribute i.
  - **Condition 2.3:** If no OIC is specified that contains the attribute i, order-preserving ciphertexts of attribute i must not be stored by CSP j. If an OIC is specified for attribute i, storing order-preserving ciphertexts of this attribute at a CSP does not give the CSP the ability to view values for attribute i. This is due to the fact that the OIC guarantees that the CSP cannot gain information from order-preserving ciphertexts of the attribute i.

Indistinguishable ciphertexts may be stored by every CSP. No information on the attribute values can be extracted by the CSP based on indistinguishable ciphertexts (cf. Section 3.4).

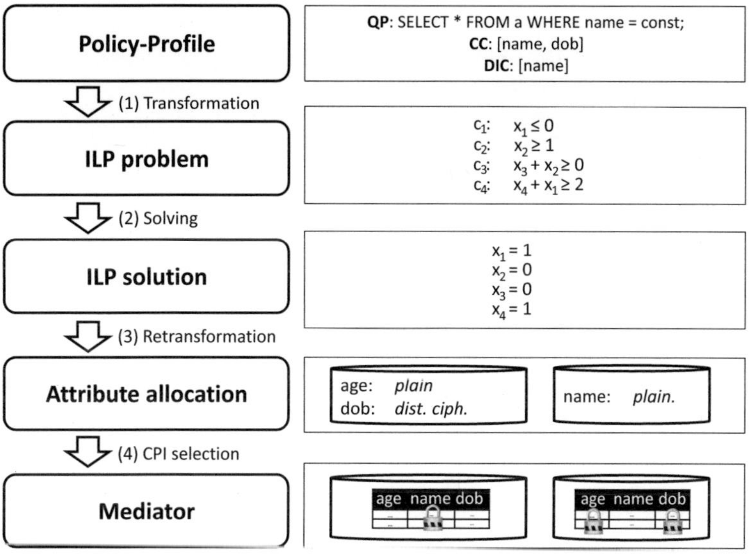

Figure 3.8: Securus' policy transformation process to generate a mediator that satisfies a given policy profile.

3.5.4 Policy Profile Transformation

Up to now, we showed how CPIs can be applied by a mediator to satisfy a policy profile. As a tunable approach, Securus must not require the user to manually choose these CPIs so that the conditions of the policy profile are satisfied and has to be able to automatically generate a mediator that satisfies the policy profile. To satisfy policy profiles, an attribute allocation needs to be determined that satisfies the conditions listed in Section 3.5.3. We denote an attribute allocation that does not violate any CCs and enables the implementation of CPIs that support the defined QPs as *feasible*. The CPIs that have to be applied at each CSP can be determined based on a feasible attribute allocation and the QPs using the CPI selection table shown in Table 3.5. Securus implements these CPIs in the mediator that is used to outsource and query the data. Determining a suitable CPI combination for a set of QPs and a given attribute allocation is straightforward. Finding a feasible attribute allocation that satisfies a policy profile, on the other hand, is hard.

Theorem 1. *Finding a feasible attribute allocation that satisfies a policy profile is NP-hard.*

Proof. See Appendix A.1 □

Securus models the problem of finding a suitable attribute allocation as an Integer Linear Programming (ILP) problem. Intuitively, the possible attribute allocations are represented by the free variables of the ILP problem and the conditions that originate from the policy profile are modeled via ILP constraints. An overview of the policy transformation process is shown in Figure 3.8. First, the requirements of the policy profile are mapped on constraints of an ILP problem (1). The resulting ILP problem is solved by an existing ILP solver like lp_solve[5] or Gurobi[6] (2). Based on the solution of the ILP problem an attribute allocation is derived (3). Based on this attribute allocation and the QPs that have to be satisfied, CPI approaches are chosen according to the CPI selection table shown in Table 3.5 and implemented in the mediator (4).

In the following, we first show how the ILP problem can be constructed to find feasible attribute allocation. Then, we show how it can be determined based on the target function of the ILP problem which of the feasible attribute allocations is the most efficient.

Finding feasible attribute allocations

In the following we provide an outline on how Securus uses ILP constraints to represent the conditions that an attribute allocation has to comply with in order to satisfy a policy profile. A full listing of the ILP problem with a detailed explanation can be found in Appendix A.2. For convenience, a legend of the used variables is shown in Figure 3.9. Unless stated otherwise all free variables of the ILP problem are binary variables, i.e., they can either take the value 1 or 0.

Mapping of Condition 1:

> *"For each QP at least one CSP has to store the relevant attributes in a representation that allows the CSP to execute efficiently the queries which are modeled by the QP."*

The free binary variable $Q_{q,j}$ denotes that CSP j is able to execute queries that match QP q. Thus, the following constraint has to hold true:

$$\forall q \in QP : \sum_{j \in S} Q_{q,j} \geq 1 \tag{3.1}$$

> *"An overview of attribute representations that are required to evaluate different query components is shown in Table 3.5. The ✓ symbol indicates that there exists a CPI based on which the CSP can evaluate the according query component and that implies the according attribute representation."*

We show how the ILP constraints for range selections (Type 2 in Table 3.5) are specified in the following. The ILP constraints for the other query component types listed in Table 3.5 can be similarly defined and are shown in Appendix A.2.

Plaintext ($P_{i,j} = 1$) or order-preserving ciphertexts ($O_{i,j} = 1$) have to be stored by CSP j to enable the evaluation of queries that contain range selections on attribute i ($i \in r(q)$). This condition can be modeled by the following constraint in the ILP problem:

$$\forall q \in QP, \forall j \in S, \forall i \in r(q) : P_{i,j} + O_{i,j} - Q_{q,j} \geq 0 \tag{3.2}$$

[5] http://lpsolve.sourceforge.net/ [last visited on March 2015]

[6] http://www.gurobi.com/ [last visited on March 2015]

Constants:
A: Set of attributes
CC: Set of confidentiality constraints (CCs)
OIC: Set of order-preserving inference constraints (OICs)
DIC: Set of distinguishable inference constraints (DICs)
QP: Set of query policies (QPs)
S: Set of storage providers (CSPs)

Free Variables:
$P_{i,j}$: Values of attribute i are stored at CSP j in plaintext.
$O_{i,j}$: Values of attribute i are stored at CSP j as order-preserving ciphertexts.
$D_{i,j}$: Values of attribute i are stored at CSP j as distinguishable ciphertexts.
$I_{i,j}$: Values of attribute i are stored at CSP j as indistinguishable ciphertexts.
$Z_{i,j}$: Values of attribute i are revealable by CSP j.
$Q_{q,j}$: CSP j can evaluate queries of QP q
$o_{i,q}$: QP q is evaluated based on order-preserving ciphertexts of attribute i.
$d_{i,q}$: QP q is evaluated based on distinguishable ciphertexts of attribute i.
$i_{i,q}$: QP q is evaluated based on indistinguishable ciphertexts of attribute i.

Functions:

$$o(i,j) = \begin{cases} O_{i,j} \text{ if } i \notin OIC \\ 0 \text{ else} \end{cases}$$

$$d(i,j) = \begin{cases} D_{i,j} \text{ if } i \notin DIC \cup OIC \\ 0 \text{ else} \end{cases}$$

$r(q)$: Set of attributes contained in range selections in QP q (Type 2 query component)

Figure 3.9: Free variables and functions of the Securus ILP problem. For a full listing of the ILP problem see Appendix A.2.

The constraint works as follows. Lets assume that $Q_{q,j} = 1$ holds, i.e., a CSP j is able to execute queries that match QP q. The equation $P_{i,j} + O_{i,j} - Q_{q,j} \geq 0$ only holds true if $P_{i,j} = 1$ or $O_{i,j} = 1$, i.e., if the CSP is allowed to store plaintext values or order-preserving ciphertexts for each attribute i on which range selections have to be evaluated ($i \in r(q)$).

Mapping of Condition 2:

> *"All CCs have to be satisfied. A CC is considered satisfied if no CSP is able to view values for all attributes that are contained in the CC."*

The free binary variable $Z_{i,j}$ denotes that CSP j can reveal attribute i. In order to satisfy a CC $c = \{a_1, \ldots, a_k\}$ that contains the attributes $a_1, \ldots, a_k$, each CSP j must not be able to reveal at least one attribute i that is included in the CC. This condition can be expressed by the following constraint in the ILP problem:

$$\forall c \in CC, \forall j \in S : \sum_{i \in c} Z_{i,j} < |c| \tag{3.3}$$

> *"The following conditions have to be satisfied to prevent a CSP j from viewing values of attribute i.*
>
> - ***Condition 2.1:*** *Plaintext values of attribute i must not be stored by CSP j.*
> - ***Condition 2.2:*** *If no DIC is specified that contains the attribute i, distinguishable ciphertexts of attribute i must not be stored by CSP j.*
> - ***Condition 2.3:*** *If no OIC is specified that contains the attribute i, order-preserving ciphertexts of attribute i must not be stored by CSP j."*

A CSP can reveal an attribute i's values ($Z_{i,j} = 1$) if it stores attribute i as plaintext ($P_{i,j} = 1$), as order-preserving ciphertexts without an OIC being specified for the i attribute ($o(i, j) = 1$) or as distinguishable ciphertexts without a DIC or an OIC being specified for the i attribute ($d(i, j) = 1$). This condition can be expressed by the following constraint in the ILP problem:

$$\forall j \in S, \forall i \in A : P_{i,j} + o(i, j) + d(i, j) - Z_{i,j} \leq 0 \tag{3.4}$$

The constraint works as follows. Lets assume that $P_{i,j} = 1$ holds, i.e., a CSP j is allowed to store plaintext values of attribute i. As all free variables cannot take negative values, $Z_{i,j}$ has to be 1 for $P_{i,j} + o(i, j) + d(i, j) - Z_{i,j}$ to be true. The argument works analogously for order-preserving ciphertexts without an OIC being specified for attribute i ($o(i, j) = 1$) or for distinguishable ciphertexts without a DIC or an OIC being specified for attribute i ($d(i, j) = 1$).

Finding optimized attribute allocations

We denote attribute allocations as *feasible* if they satisfy the conditions of the policy profile. In many cases, multiple feasible attribute allocations exist that satisfy a policy profile. With regard to efficiency, a feasible solution is not necessarily *optimal*. We show that considerable efficiency differences can exist between feasible and optimal solutions in Section 3.5.6.

> *Example:* For a policy profile that contains the CC `[A,B]` and the QP `SELECT SUM(A) ...`, both attribute allocations "CSP 1 may store A in plaintext and B as indistinguishable ciphertexts" and "CSP 1 may store B in plaintext and A as indistinguishable ciphertexts" are feasible. However, while queries of the QP can be evaluated by CSP 1 based on plaintext values with the first attribute allocation, the second attribute requires to apply partially homomorphic encryption on A. Thus, the second attribute allocation can be considered less efficient.

Securus enables to determine optimized solutions by making use of the ILP problem's target function. In this chapter, we propose a target function that takes into account query latency and transmission overhead. The concept can be extended to also consider other efficiency metrics such as storage costs (see Appendix A.6).

Query latency costs: Let $c^o_{q,i}$, $c^d_{q,i}$, and $c^i_{q,i}$ be the average latency costs in milliseconds to evaluate queries of QP q on order-preserving, distinguishable, and indistinguishable ciphertexts of attribute i respectively. These costs are induced by the CPIs that are implemented in the mediator which are chosen based on the attribute allocation and the QPs. Furthermore, $o_{i,q}$, $d_{i,q}$, and $i_{i,q}$ are free binary variables that state whether queries of QP q are evaluated on order-preserving ($o_{i,q} = 1$), distinguishable ($d_{i,q} = 1$), or indistinguishable ciphertexts ($i_{i,q} = 1$). In the ILP problem, $o_{i,q}$, $d_{i,q}$, and $i_{i,q}$ are free variables and $c^o_{q,i}$, $c^d_{q,i}$, and $c^i_{q,i}$ are constants.

The query latency cost can be determined by summing up the latency costs induced for each QP for every attribute i that is relevant for the QP ($i \in q$) as shown in Equation 3.5:

$$\sum_{q \in QP} \sum_{i \in q} (c^o_{q,i} \cdot o_{i,q} + c^d_{q,i} \cdot d_{i,q} + c^i_{q,i} \cdot i_{i,q}) \tag{3.5}$$

Notice that Equation 3.5 expresses the query latency costs that are induced if for each specified QP the same number of queries are contained in the query workload. It is straightforward to extend Equation 3.5 to also consider unevenly distributed query workloads. To do that, for each QP the relative occurrence frequency of workload queries that match the QP has to be applied as weight on the induced query latencies. The occurrence frequencies can be specified together with the QP in the policy profile.

We make the assumption that the latency cost that is induced by each attribute representation (and therefore of utilized CPIs) is independent from the representation of (and therefore the used CPIs on) other attributes. Based on this assumption the costs can be added up as shown in Equation 3.5. We discuss this assumption in Section 3.5.7.

The allocations of the free variables $o_{i,q}$, $d_{i,q}$, and $i_{i,q}$ are subject to constraints. A query of QP q has to be evaluated based on order-preserving ciphertexts of attribute i, i.e., $o_{i,q} = 1$), if the CSP j that is used to evaluate the query ($Q_{q,j} = 1$) may only store order-preserving ciphertexts ($O_{i,j} = 1$). The same holds true for distinguishable and indistinguishable ciphertexts. These constraints are specified in the ILP problem by Securus as shown in Equation 3.6-3.8:

$$\forall q \in QP, \forall i \in q, \forall j \in S: \quad o_{i,q} - Q_{q,j} - O_{i,j} \geq -1 \tag{3.6}$$

$$\forall q \in QP, \forall i \in q, \forall j \in S: \quad d_{i,q} - Q_{q,j} - D_{i,j} \geq -1 \tag{3.7}$$

$$\forall q \in QP, \forall i \in q, \forall j \in S: \quad i_{i,q} - Q_{q,j} - I_{i,j} \geq -1 \tag{3.8}$$

Providing performance models for each CPI to determine the constants $c^o_{q,i}$, $c^d_{q,i}$, and $c^i_{q,i}$ is not within the scope of this thesis. However, we will show in Section 3.5.6 that big performance differences between CPIs exist and optimizing the usage of CPIs to satisfy a policy profile can improve performance drastically.

Transmission cost: Let $t^o_{q,i}$, $t^d_{q,i}$, and $t^i_{q,i}$ be the average transmission costs in bytes that occur when a query of QP q is evaluated on order-preserving, distinguishable, and indistinguishable ciphertexts of attribute i respectively. The transmission costs are induced by the CPIs that are implemented in the mediator based on the attribute allocation and the QPs.

As shown in Equation 3.9 the transmission cost can be determined by summing up the transmission costs induced for each QP for every attribute i that is relevant to the QP ($i \in q$):

$$\sum_{q \in QP} \sum_{i \in q} (t^o_{q,i} \cdot o_{i,q} + t^d_{q,i} \cdot d_{i,q} + t^i_{q,i} \cdot i_{i,q}) \tag{3.9}$$

To support query workloads that are unevenly distributed in the sense that they contain a lot of queries that match QP q_1 and only few queries that match QP q_2, the transmission costs can be weighted like we explained for the query latency model. The following example illustrates how to determine the transmission costs for a given query.

> *Example:* Consider the following query q: `SELECT b ... WHERE a=?`. Regardless of whether q is evaluated on order-preserving or distinguishable ciphertexts of attribute a and b, the same CPIs are applied (see Table 3.5). The equality selection `a=?` is evaluated on hash values in both cases. Furthermore, probabilistic encryption (e.g., AES) is applied on attribute b values. Thus, the transmission costs $t^o_{q,'a'}$ and $t^d_{q,'a'}$ both amount to the size of a hash value. The transmission costs $t^o_{q,'b'}$ and $t^d_{q,'b'}$ both amount the size of an AES ciphertext on attribute b times the expected number of records that match the query. The constant $t^i_{q,'a'}$ does not need to be set, since the equality selection `a=?` cannot be evaluated on indistinguishable ciphertexts. Thus, $i_{q,'a'} = 0$ holds in any case and the value of $t^i_{q,'a'}$ is irrelevant for the optimization problem. In fact, the

$t^i_{q,'a'}$ variable can be omitted entirely in the ILP problem. By pruning the ILP problem in such a way, the efficiency of the policy transformation process can be optimized in future work.

Weighted target function: The query latency cost and the transmission cost of an attribute allocation can be determined based on Equation 3.5 and Equation 3.9. In order to jointly minimize both query latency and transmission cost, Equation 3.5 and Equation 3.9 can be weighted with factors w_l and w_t depending on whether query latency is valued more than transmission cost or vice versa in the deployment scenario. The resulting function is shown in Equation 3.10 and constitutes the target function that has to be minimized in the ILP problem specified by Securus.

$$w_l \cdot \sum_{q \in QP} \sum_{i \in q} (c^o_{q,i} \cdot o_{i,q} + c^d_{q,i} \cdot d_{i,q} + c^i_{q,i} \cdot i_{i,q}) + w_t \cdot \sum_{q \in QP} \sum_{i \in q} (t^o_{q,i} \cdot o_{i,q} + t^d_{q,i} \cdot d_{i,q} + t^i_{q,i} \cdot i_{i,q}) \quad (3.10)$$

3.5.5 Policy Conflict Detection and Resolution

It is possible to specify unsatisfiable PPs that contain policy conflicts. As a trivial example, consider the PP shown in Figure 3.10 that contains a QP `SELECT ... WHERE age < ? AND name LIKE ?` and a CC `[name]`. To enable a CSP to evaluate the QP efficiently, the CSP needs to store plaintext or order-preserving ciphertexts of the attribute *name* (see Table 3.5). However, this violates the CC `[name]` as no OIC is specified for *name* and the CSP is not allowed to store neither plaintext nor order-preserving ciphertexts.

If policies within a PP are conflicting, the solution space of the ILP problem is empty. Thus, the ILP problem is unsolveable and the authors of the PP are notified. Manually identifying conflicting policies in PPs that contain many policies can be hard due to the size of the policy profile and the complexity of locating policy conflicts. In particular, the NP-hardness proof which we provided for the problem of finding an attribute allocation that satisfies a policy profile implies that finding policy conflicts is also NP-hard. However, in many cases only a small subset of the policies are in conflict with each other. To reduce the complexity that comes with conflict resolution, it is worthwhile to isolate conflicting policies from policies that do not stand in conflict with other policies. In the following we will show how policy conflicts can be *isolated*, i.e., how the conflicting policies of a PP can be determined, and how policy conflicts can be *resolved*.

Policy Conflict Isolation

In most cases, only a small subset of policies in the PP are in conflict with each other. For instance, in the PP shown in Figure 3.10 only the two underlined policies are conflicting. This conflict is independent from the other policies in the sense that even if the PP contained only the two underlined policies, it would be unsatisfiable.

As the constraints of the ILP problem that is formulated by Securus are bijectively mapped to the policies in the PP (cf. Section 3.5.4), a PP that is unsatisfiable due to conflicting policies inherently results in an infeasible ILP problem. Conversely, an infeasible ILP problem implies that the PP contains policy conflicts.

```
PolicyProfile AnotherEmployeeDB{
        Tables{
                Employees{
                        name            :STRING(10)
                        address         :STRING(40)
                        age             :INTEGER
                        phoneNumber     :STRING(15)
                }
        }
        QueryPolicies{
                SELECT * FROM Employees
                WHERE
                age < ?  AND name LIKE ?;

                SELECT SUM(age) FROM Employees
                WHERE
                age < ?;
        }
        ConfidentialityConstraints{
                [name]
                [phoneNumber,address]
                [phoneNumber,age]
        }
}
```

Figure 3.10: Policy Profile with conflicts. Conflicting policies are underlined.

Modern ILP solvers like Gurobi offer the possibility to identify *irreducible inconsistent sets* (IISs) of ILP constraints. An IIS is defined to be a set of constraints that implies an empty solution space[7]. If one constraint of the IIS is removed, the solution space is not empty anymore due to the IIS. An ILP problem may contain multiple IISs just like a PP may contain multiple sets of conflicting policies.

As policies are bijectively mapped to ILP constraints, Securus can determine sets of conflicting policies within a PP as follows. Securus lets the ILP solver determine the IISs within the infeasible ILP problem. For each ILP constraint within an IIS, the policies that were mapped on this constraint are added to a set of conflicting policies. In this way, a set of conflicting policies is generated for each IIS. These sets represent the isolated policy conflicts within a policy profile.

Note that by relying on generic ILP methods, our method to isolate policy conflicts is independent from the actual policy model that is used by Securus. Thus, if the policy model of Securus is extended in future work, policy conflicts can be isolated just like in the current model without any adaptations.

Policy Conflict Resolution

Policy conflicts arise when query requirements clash with data confidentiality requirements. In fact, a policy profile that only contains QPs cannot contain any policy conflicts. Likewise, a PP that contains only CCs cannot contain policy conflicts. To resolve conflicts in PPs, either the CCs or the QPs that are conflicting have to be relaxed. For instance, in the previous example the QP `SELECT ... WHERE age < ? AND name LIKE ?` can be relaxed to `SELECT ... WHERE age < ?`. Thus, the SP can evaluate the `age < ?` condition and the mediator can evaluate the `name LIKE ?` condition on the returned results to filter out false-positive records.

There are multiple approaches on how to perform a policy relaxation to resolve the identified policy conflicts.

Manual conflict resolution can be performed by the risk manager for each policy conflict. For instance, in the previous example the risk manager can decide whether the CC `[name]` can be relaxed. If this is not the case, the risk manager has to relax the QP `SELECT ... WHERE age < ? AND name LIKE ?` to `SELECT ... WHERE age < ?` and accept the performance loss that is induced by having to evaluate the query condition on the attribute *name* at the mediator.

Deny-overrules conflict resolution can be used to automate the resolution process without involving the risk manager. As shown in the previous example, for each policy conflict Securus leaves the CCs of a policy profile untouched (deny) and relaxes the critical QPs until the policy conflict is resolved. If performance is considered more important than data confidentiality, the same concept can be applied in a *allow-overrules* fashion where the CCs are relaxed and the QPs remain untouched.

7 If the solution space is empty, no solution to the specified ILP problem exists and the ILP problem is considered infeasible.

Weighted conflict resolution can be used to automate the resolution process in case a preference can neither be expressed for confidentiality nor for query efficiency. The risk manager can annotate each policy with a *weight* value. Based on these weights, Securus can relax policies in such a way that the weights of the relaxed policies are minimized[8].

Beyond the scope of this thesis, yet a promising future research direction is to investigate how the relaxation of QPs can be optimized in such a way that the performance impact of partially evaluating the query at the mediator is minimized. A good starting point would be to build on approaches of the database community that optimize query execution in conventional databases [Cha98, CN07].

3.5.6 Tunability of Securus

CPI Efficiency and Optimization Potential of Securus

To evaluate the efficiency benefits that Securus provides, we conducted measurements based on the TPC-C benchmark [Tra10]. The TPC-C Benchmark portrays the activities of a wholesale supplier and is designed to be "representative of complex [online transaction processing] (OLTP) application environments" [Tra10]. The benchmark models a database into which orders are entered, payments are recorded, orders are checked, and the level of stock at the warehouses is monitored. The database contains 100000 item types, 30000 customers, more than 300000 orders. The TPC-C benchmark is widely used to evaluate the performance of databases. We measured queries on the `ORDERLINE` table which in particular contains the attributes *O* (`OL_O_ID`, the order ID), *D* (`OL_D_ID`, the ID of the order's district), *W* (`OL_W_ID`, the ID of the order's warehouse), *A* (`OL_AMOUNT`, the amount ordered) and *I* (`OL_I_ID`, the ID of the item ordered). The ORDERLINE table contains around 300000 records. We measured the latency of the queries `SELECT SUM(A) FROM ORDERLINE WHERE W=?, D=?, O=?` and `SELECT A,I FROM ORDERLINE WHERE W=?, D=?, O=?` which are part of the workload that is defined by the TPC-C benchmark.

The results of our measurements for the two types of queries are shown in Figure 3.11 and Figure 3.12 respectively for a varying set of applied CPI approaches. We measured the latencies of 2000 queries to calculate the average query latency and the 95% confidence intervals for each query and each set of CPI approaches.

Differences in CPI efficiency: Whereas a full performance evaluation of all existing CPIs is beyond the scope of this thesis, the conducted measurements show that substantial performance differences exist between CPIs. For instance, Figure 3.11 shows that evaluating equality selections based on hash values rather than plaintext values only induces a slight overhead ("A,W,D,O plaintext" vs. "A plaintext; W,D,O hashed"). Aggregating attribute values via a partially homomorphic scheme induces an overhead of close to 500% for the investigated query ("A plaintext; W,D,O hashed"

[8] This optimization problem can be easily solved by formulating it as another ILP problem and apply common ILP solvers to solve the ILP problem.

vs. "A hom.enc.; W,D,O hashed"). We used the Paillier encryption scheme [Pai99] in this case. A more detailed investigation of CPI performance can be found in [PRZB11] for selected CPIs.

Efficiency gains and losses via data fragmentation: Fragmenting data into multiple unlinkable fragments can yield efficiency benefits as the measurements presented in Figure 3.11 show. For instance, if a confidentiality constraint `[A,X]` is satisfied by storing X in another fragment than A, W, D and O, neither X nor A have to be protected. Thus, the high efficiency overhead for applying partially homomorphic encryption on A can be avoided.

However, not including attributes that are necessary to evaluate the conditions of the same query in a fragment can lead to severe efficiency losses as the measurements presented in Figure 3.13 show. We measured the query latencies that are induced for executing the query `SELECT A,I FROM ORDERLINE WHERE W=?, D=?, O=?` based on fragments that all contain the attributes A and I. The fragments differ with regard to which subset of the attributes W, D and O is contained. In Figure 3.13, we mark the fragments that contain W, D and O with "`... WHERE W,D,O`", the fragments that just contain W and D with "`... WHERE W,D`", and so on.

The efficiency losses result from the fact that the CSP can only evaluate query conditions that refer to attributes that are contained in the fragment on which the query is executed on. Conditions on attributes that are not contained in the fragment can only be evaluated after the records are decrypted by the mediator. Thus, false-positive records that do not match the query have to be transmitted to the mediator. As Figure 3.13 illustrates, the amount of false-positive records that have to be transmitted depends on the fragmentation of the data and the selectivity of each attribute, i.e., the fraction of records that match the query condition.

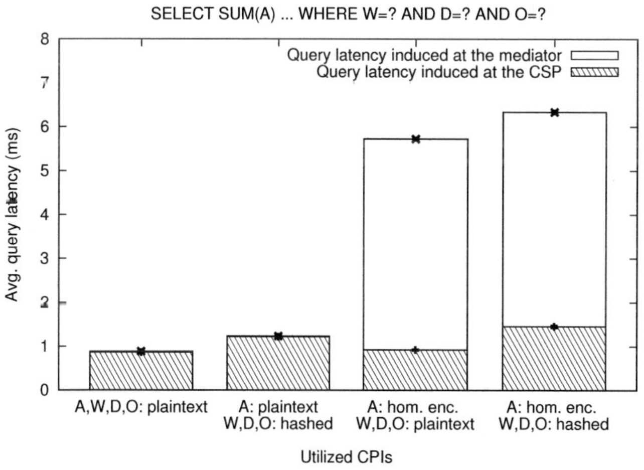

Figure 3.11: Average query latencies for different CPI combinations of the TPC-C query `SELECT SUM(A) FROM ORDERLINE WHERE W=?, D=?, O=?`.

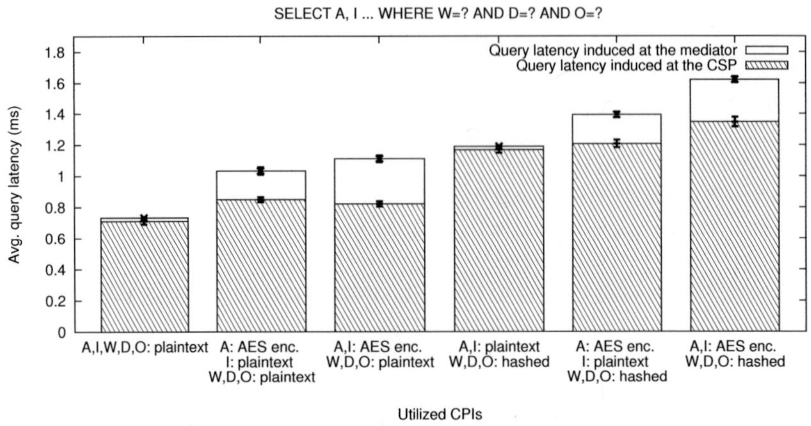

Figure 3.12: Average query latencies of the TPC-C query `SELECT A,I FROM ORDERLINE WHERE W=?, D=?, O=?` for different CPI combinations.

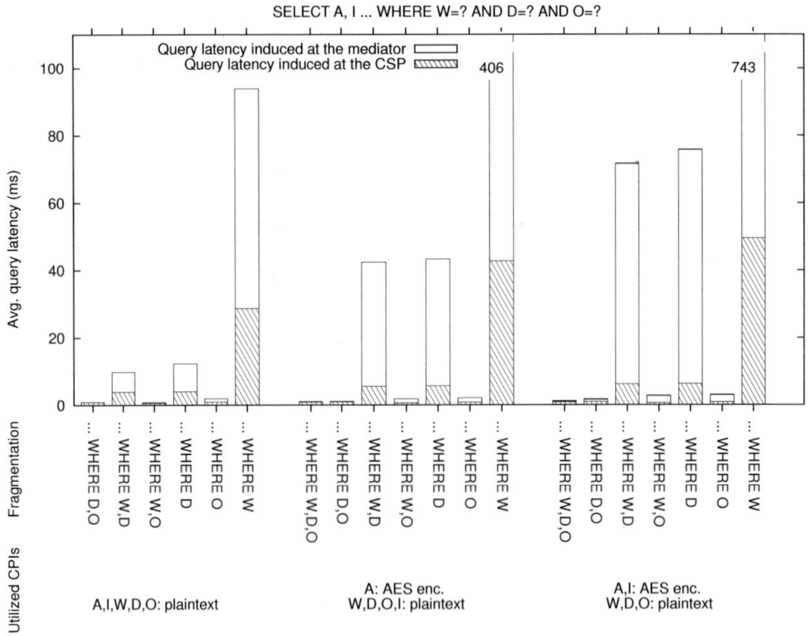

Figure 3.13: Average query latencies of the TPC-C query `SELECT A, I FROM ORDERLINE WHERE W=?, D=?, O=?` for different fragments and CPIs. Fragments that contain the *W*, *D* and *O* attributes are marked with "`. . . WHERE W,D,O`". Fragments that only contain *W* and *D* are marked with "`. . . WHERE W,D`".

For instance, if only W is contained in the fragment, only the query's condition on W can be evaluated by the CSP and all matching records have to be transferred to the mediator where they are decrypted and false-positive records are discarded based on the conditions on D and O. When comparing the measurements for `... WHERE W` with `... WHERE W,D,O` in Figure 3.13, it becomes apparent that transmitting and decrypting false-positive records can be very expensive with regard to efficiency.

Securus does partition data on multiple CSPs but not combinations of attributes that are required to evaluate the conditions of a query. Therefore, it avoids efficiency losses that occur due to data partitioning. In future work, once more detailed performance models are available to predict the overheads induced by data fragmentation, the ILP problem formulation of Securus can be extended to partition attributes that are required to evaluate a query if the efficiency costs can be expected to be lower than those of applying CPIs.

Tunability of Securus: Securus aims to find the optimal combination of CPIs that satisfies the specified policy profile and minimizes the expected efficiency cost. The stricter the confidentiality requirements are and the more background knowledge the assumed attackers possess, the more efficiency cost is induced by CPIs that have to be utilized to satisfy the policy profile and, thus, the less efficient the optimal outsourcing solution can be. This is illustrated in Figure 3.14 which shows the performance of Securus mediators that satisfy different policy profiles for the two investigated queries. Note that the green solutions constitute optimal solutions, i.e., based on the CPIs available to Securus, no more efficient CPI sets can be found that satisfy the policy profile.

Trade-off between confidentiality requirements and efficiency: If no CCs are specified in the policy profile (1 in Figure 3.14), all attributes can be outsourced in plaintext. Thus, the queries Q_1 and Q_2 can be evaluated on plaintext attribute values with an average query latency of 0.7 ms and 0.9 ms, respectively. If CCs are specified for each single attribute (2 in Figure 3.14), all attributes have to be protected. As DICs are specified for the attributes W, D, and O, the conditions of Q_1 and Q_2 can be evaluated based on hash values. The attribute values of A and I have to be probabilistically encrypted to protect them. Based on this CPI combination, the average latency for executing Q_1 and Q_2 queries amounts to 1.6 ms and 6.4 ms, respectively. Thus, by specifying the CCs, the query latency of Q_1 is increased to ~133% and the query latency of Q_2 is increased to ~492% compared to the mediator (1) that satisfies the policy profile with no specified CCs.

Efficiency optimization for given confidentiality requirements: If it is not required that each attribute has to remain confidential, but that the attribute combinations `[A,W]`, `[A,D]`, and `[A,O]` have to be protected, more efficient CPI combinations can be found (3 in Figure 3.14). To satisfy the CCs, it suffices to protect either A or W, D, and O at the CSPs. Protecting W, D, and O implies that the equality selections of Q_1 and Q_2 have to be evaluated on hashes rather than plaintext records (green rectangles in Figure 3.14 (3)). Protecting attribute A implies that the mediator has to decrypt probabilistically encrypted values of A when evaluating queries of QP Q_1 and partially homomorphic encryption has to be applied to aggregate values of attribute A in order to evaluate queries of QP Q_2 (red rectangles in Figure 3.14 (3)).

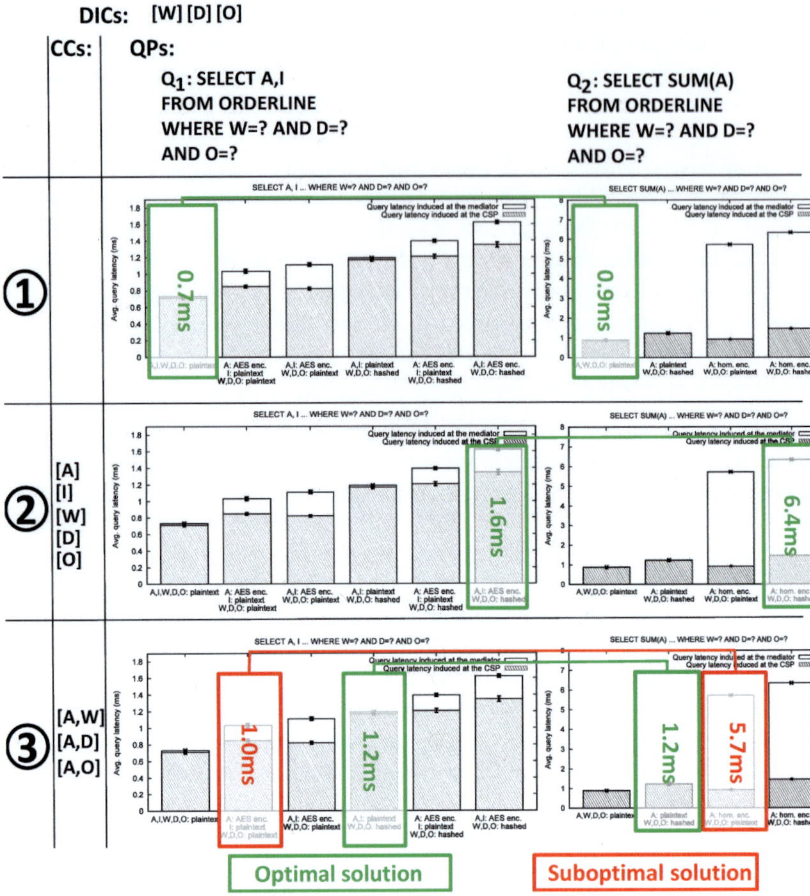

Figure 3.14: Example to illustrate the trade-off between confidentiality requirements and efficiency with regard to query latency. The example is based on our TPC-C query latency measurements that are shown in Figure 3.11 and 3.12. For policy profiles with different confidentiality constraints (1, 2, and 3), the query latency that is induced by the optimal CPI choice is highlighted green. The query latency that can be achieved based on the optimal CPI choice improves if less confidentiality constraints are specified (2 vs. 1) or if the confidentiality constraints leave room to choose which attributes have to be protected (2 vs. 3).

Protecting attribute A can be considered more efficient with regard to QP Q_1 (1.0 ms vs. 1.2 ms). However, it implies severe efficiency costs when evaluating queries of QP Q_2 (0.9 ms vs. 6.4 ms). Thus, if only one CSP is available, it is more efficient to protect the attributes W, D, and O instead of attribute A.

From the CPI combinations that satisfy the policy profile, Securus chooses the optimal one with regard to efficiency (cf. Section 3.5.4). Note that Securus does not aim to minimize the number of protected attributes, but to minimize the efficiency cost. For instance, as shown in Figure 3.14 (3), it is more beneficial to protect multiple attributes W, D, and O to avoid the high efficiency costs of applying partially homomorphic encryption to protect the single attribute A.

Securus finds solutions for policy profiles that are as efficient as possible and still satisfy the confidentiality requirements. The measurements in Figure 3.14 show that this can yield big efficiency benefits even in simple scenarios.

Interdependencies between CPI query latency overheads: As we showed in Section 3.5.4, in the target function of the ILP problem Securus makes the assumption that costs induced by a CPI are independent from the other applied CPIs. Our measurements indicate that the utilized CPIs do not majorly affect each other's induced query latency overhead. For instance, consider Figure 3.12. The induced cost for hashing attributes W, D, and O ("W,D,O: hashed" instead of "W,D,O: plaintext") amounts to 0.4 ms to 0.5 ms regardless of the other applied CPIs. Consider, for instance, the measurements of the "A,I,W,D,O: plaintext" mediator (0.75 ms) vs. the measurements of the "A,I: plaintext, W,D,O: hashed" mediator (1.2 ms). Other examples include "A: AES enc, I: plaintext, W,D,O: plaintext" (1.0 ms) vs. "A: AES enc, I: plaintext, W,D,O: hashed" (1.4 ms) as well as "A,I: AES enc, W,D,O: plaintext" (1.1 ms) vs. "A,I: AES enc, W,D,O: hashed" (1.6 ms). The seemingly small deviation (0.4 ms vs. 0.5 ms) can be explained by the confidence intervals.

Notice that Securus chooses an optimal set of CPIs in the measured scenario even if the CPI query latency overheads are interdependent and the assumed CPI efficiency overheads deviate from the actual overheads that are induced by each CPI to a small degree. For instance, in Figure 3.12 the costs of hashing W, D, and O are in the range of 0.4 ms to 0.5 ms. The costs of encrypting attribute A based on AES are in the range of 0.2 ms to 0.25 ms. Thus, in the worst case Securus would assume that $c^d_{q_1,'W'} + c^d_{q_1,'D'} + c^d_{q_1,'O'} = 0.4$ and $c^i_{q_1,'A'} = 0.25$. Even in this case Securus would make the optimal decision to encrypt attribute A ("A: AES enc, I: plaintext, W,D,O: plaintext") instead of hashing attributes W, D, and O ("A,I: plaintext, W,D,O: hashed") to satisfy the CCs `[A,W]`, `[A,D]`, and `[A,O]`.

Comparison of Securus to Other Approaches

In the following we illustrate the advantages that originate from the tunability of Securus. To achieve that, we compare other approaches to Securus based on the exemplary policy profile shown in Figure 3.15 and our TPC-C benchmark measurements.

To satisfy the CC `[X,A]`, Securus stores attribute X on CSP 1 to avoid having to protect A at CSP 2 which would imply severe efficiency overheads to evaluate queries Q_2 since A would have to be aggregated based on partial homomorphic encryption.

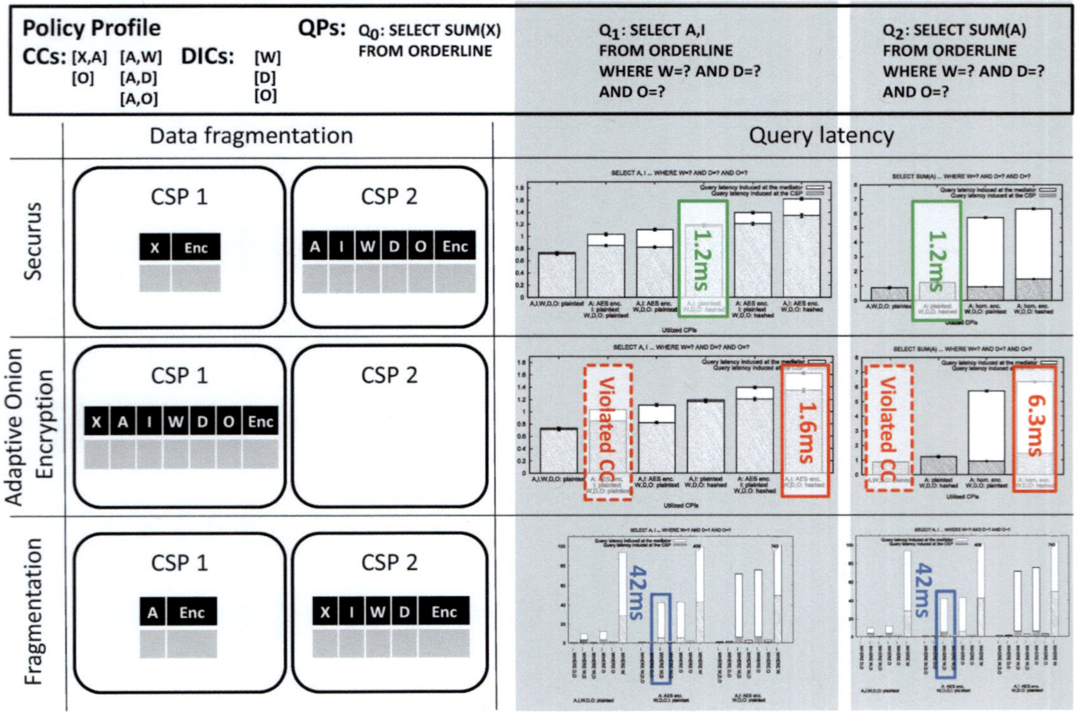

Figure 3.15: Example to illustrate the efficiency benefits of Securus compared to other approaches. The example is based on our TPC-C query measurements shown in Figure 3.11, 3.12, and 3.13. The highlighted query latencies are induced by the database outsourcing approaches that enforce the listed policy profile. Securus adapts the CPI choice to the policy profile and is thus able to optimize query latency as much as possible. This is not the case for the other approaches.

Furthermore, to satisfy the CCs `[A,W]`, `[A,D]`, `[A,O]`, and `[O]`, the attributes *W*, *D*, and *O* have to be protected at CSP 2 so that CSP 2 may store plaintexts of *A*. The specified DICs `[W]`, `[D]`, and `[O]` state that the attributes can be considered protected if only hashes of *W*, *D* and *O* are outsourced. The query latencies that are induced by this CPI combination are highlighted in green in Figure 3.15. Note that the best shown query latencies cannot be reached by any approach as they require the attributes *A*, *I*, *W*, *D*, and *O* to be available as plaintext at one CSP which conflicts with the CCs `[A,W]`, `[A,D]`, `[A,O]`, and `[O]`.

Adaptive onion encryption approaches [GHH+14, KHK+13] such as CryptDB [PRZB11, PZB11] do not consider CCs and QPs but aim to keep the data as protected as possible while still being able to evaluate all queries "efficiently" (see Section 3.3). Furthermore, they to not make use of fragmentation to hide attribute combinations. In Figure 3.15, to evaluate the conditions on *W*, *D*, and *O* of the queries Q_1 and Q_2, the CSP needs access to distinguishable ciphertexts of the attributes such as hash values. To execute queries of the QP Q_1, no plaintext values of *A* and *I* are necessary and it suffices that probabilistically encrypted values are stored at CSP 1. To aggregate attribute values of *A* in query Q_2, no plaintext values of *A* are necessary. Partially homomorphic encrypted values can be summed up by CSP 1 and the result can be decrypted at the mediator. The query latencies that are induced by these CPI combinations are highlighted in red in Figure 3.15. Especially the unnecessary homomorphic aggregation of attribute *A* constitutes a big efficiency overhead compared to the Securus solution that also satisfies the confidentiality requirements that are specified in the policy profile. Thus, compared to approaches such as CryptDB, Securus optimizes the efficiency of queries as much as possible by not protecting attributes unnecessarily.

If other previously executed queries required the onion encryption approach to decrypt all values to plaintext, the queries Q_1 and Q_2 can be evaluated on plaintext values and the query efficiency increases (see the red dotted lines in Figure 3.15). However, as CSP 1 stores plaintext values of all attributes, the specified CCs would be violated. Thus, one disadvantage of adaptive onion encryption approaches is that the user has no opportunity to explicitly specify the data's confidentiality requirements, but rather has to estimate the loss of confidentiality that is induced by executing a given query.

Securus' advantages over onion encryption approaches include the following:

- Securus provides fixed confidentiality guarantees.
- Securus tunes the confidentiality of an outsourcing solution to satisfy the deployment scenario requirements and optimizes efficiency at the same time[9].
- Securus leverages fragmentation for additional efficiency benefits.

9 Approaches such as CryptDB could potentially be extended by adding a policy layer that enforces the decryption of onions as far as possible when evaluating a query and prevents onions from being decrypted if this would violate CCs. However, in most cases this approach will not result in the optimal solution. For instance, attributes that have been decrypted but are only needed to evaluate a single query may prevent the decryption of other attributes that are relevant for the majority of future queries.

Fragmentation approaches [DCdVFJ+14, For11, HGSB13, AGH11, ABG+05] aim to split the data in multiple fragments to prevent that attributes of a CC can be linked. Whether an attribute may be present in multiple fragments depends on whether the fragments are stored at a single CSP or multiple non-colluding CSPs. Fragments that are stored by a single CSP can be possibly linked via attributes that they have in common. If a CC contains just one attribute, this attribute has to be encrypted. When executing a query, a query optimizer has to choose the fragment that is suited best with regard to efficiency. Some approaches even consider the query workload when creating the fragmentation.

In Figure 3.15, to satisfy the CCs `[X,A]`, `[A,W]`, `[A,D]`, and `[A,O]`, attribute *A* has to be stored in a different fragment than the attributes *X*, *W*, and *D*. The attribute *O* must not be stored in plaintext in any fragment due to the CC `[O]`.

The query latency that is induced by this fragmentation is highlighted in blue in Figure 3.15. Compared to Securus and the adaptive onion encryption approaches, a big efficiency overhead is induced (42 ms vs. 1.2 ms vs. 1.6 ms) since a lot of false-positive records have to be transmitted to the mediator. How many false-positives are transmitted depends on the distribution of the outsourced data and the quality of the query optimizer decisions.

Securus' advantages over fragmentation-based approaches:

- Securus prevents trivial linking attacks when the attacker can observe inserts[10].
- Securus does not suffer from query latency and transmission overhead that is induced by transmitting false-positive records as it only fragments data if this does not imply that false-positive records have to be transmitted to the mediator. Thus, unlike fragmentation approaches Securus' query efficiency is independent from the frequency distribution of the outsourced attribute values.

3.5.7 Evaluation and Discussion

Security

It has to be distinguished between the security of the Securus approach itself, i.e., whether Securus mediators satisfy the specified requirements in the policy profile, and the secure application of the Securus approach, i.e., whether the Securus mediator protects the data as intended by the author of the policy profile.

Security of Securus: Securus mediators correctly enforce the confidentiality requirements that are specified in the policy profile if 1) the CPIs have the security properties which they guarantee, and 2) the attackers are honest-but-curious, non-colluding and have no background knowledge on queries.

A CC is satisfied if each CSP is unable to reveal values of at least one attribute that is contained in the CC. Based on the assumption that the attackers are non-

[10] Theoretically, fragmentation approaches also support this by storing each fragment on a separate CSP. However, *n* non-colluding CSPs would be required, where *n* is the number of fragments that are necessary to satisfy the CCs. Thus, in many deployment scenarios, this approach is not applicable.

colluding and cannot infer attribute values based on observed queries via background knowledge on query patterns, this is the case if condition 2 in Section 3.5.3 holds. Condition 2 is directly mapped on constraints in the ILP problem. As the final CPI choice that is implemented in the mediator satisfies the ILP constraints, it also satisfies condition 2. If the CPIs do indeed only leak the information which they advertised to an honest-but-curious attacker, the generated Securus mediator correctly enforces the specified CC requirements.

Based on standard security definitions of encryption schemes, it can be easily shown that the CPIs that are used by Securus to date indeed only leak the information that is allowed for their attribute representation at the CSP. For instance, Securus makes use of probabilistic encryption schemes such as the symmetric encryption scheme AES and the partially homomorphic encryption scheme Paillier [Pai99] that are commonly assumed to be IND-CPA secure [KL07] (*Indistinguishability against Chosen-Plaintext Attacks*), i.e., their ciphertexts are indistinguishable from each other against an attacker who is able to produce the ciphertexts of arbitrary plaintexts. Thus, the ciphertexts are by definition indistinguishable ciphertexts.

Furthermore, Securus can use pseudorandom functions to enable the evaluation of equality selections on distinguishable ciphertexts. Pseudorandom functions can be shown to be IND-DCPA secure [BKN02] (*Indistinguishability against Distinct Chosen-Plaintext Attacks*), i.e., their ciphertexts are indistinguishable from each other as long as the same plaintext value is only encrypted once. In other words, they just leak the information on which ciphertexts correspond to the same plaintext values and which do not. Thus, the ciphertexts are by definition distinguishable ciphertexts.

Securus makes use of order-preserving encryption schemes which are assumed to be IND-OCPA secure [PLZ13] (*Indistinguishability against Ordered Chosen-Plaintext Attacks*), i.e., they reveal no information about the encrypted plaintexts beside the order. Thus, the ciphertexts are by definition order-preserving ciphertexts.

A universal composability [Can01] security proof exists to show that CPIs only leak the amount of information that is permitted by the ICs that are specified in the policy profile [NK15]. For the sake of readability, we omit the proof in this thesis and refer the interested reader to [NK15].

Secure application of Securus: Securus generates mediators that securely enforce the requirements which are specified in the policy profile. If errors are made when specifying these requirements *before* Securus is invoked, Securus will provide a secure database outsourcing solution in most cases. If the requirements do not correctly reflect the intended confidentiality requirements, the Securus mediator enforces the wrong requirements. Thus, for a mediator to be considered secure, the PP has to be correctly specified by the risk manager in the sense that the CCs and ICs correctly reflect the confidentiality requirements and the ability of the attacker to apply background knowledge. In particular, determining confidentiality requirements includes modelling semantic dependencies between attributes in the PP via CCs to avoid that other CCs are undermined. This is part of the risk management process.

Semantic dependencies within the outsourced database constitutes a problem that is not specific to Securus but to all outsourcing approaches that selectively apply encryption techniques and avoid encrypting the whole database. Such dependencies are called *functional dependencies* [DCdVFJ+14]. A functional dependency between attributes A and B exist if it is possible to derive some of attribute B's values from attribute A's values. If the risk manager is aware of functional dependencies she can model them with CCs so that Securus can cope with them. To do so, each CC that contains attribute A has to be duplicated and A has to be replaced by B in the copy.

> *Example:* Consider a database with the attributes *name, city, ZIP* and the requirement that *name* must not be mappable to a city that is expressed by the CC `[name,city]`. Enforcing the CC does not satisfy the original requirement: Protecting the attribute *city* serves no purpose if the attribute *ZIP* is not protected, since each ZIP code can be mapped to a city. To cope with the functional dependency between *ZIP* and *city*, another CC `[name,ZIP]` has to be defined. Thus, either *name* or *city and ZIP* are protected and the functional dependency cannot be exploited.

Based on existing CCs and known functional dependencies, the task of deriving additional CCs that enforce the intention behind the original CCs in face of functional dependencies can be easily automated. Determining methodologies to identify functional dependencies is an interesting research topic for future research.

Optimality

In the following we evaluate whether Securus produces efficiency-optimal mediators for policy profiles with regard to the induced query latency and transmission cost. By definition, the solution to an ILP problem is optimal with regard to the target function. Thus, the solution that Securus derives from the ILP solution is also optimal if the ILP problem correctly reflects the problem of finding an optimal solution to a policy profile. If the ILP problem is correctly modeled, Securus produces optimal solutions.

The ILP problem is correctly modeled if the constraints are correctly modeled (feasibility) and the target function is correctly modeled (optimality). The ILP constraints are correctly modeled if they reflect the conditions that a feasible solution for a police profile has to satisfy (see Section 3.5.3). This is the case as each condition maps to a set of ILP constraints that enforce the condition.

The target function is correctly modeled if the following assumptions hold: 1) the transmission/latency costs $c^o_{q,i}$, $c^d_{q,i}$, $c^i_{q,i}$ and $t^o_{q,i}$, $t^d_{q,i}$, $t^i_{q,i}$ are correctly specified by the user and 2) the latency costs of each CPI are independent from other utilized CPIs.

The latency/transmission costs for the given deployment scenario can be determined by conducting careful measurements. The assumption that the query latency overhead of a CPI is completely independent from that of other CPIs may not hold in all deployment scenarios. Coping with inter-dependent CPI query latency overheads implies a combinatorial explosion. Thus, modeling inter-dependent query latency overheads in the ILP is possible. However, the approach would not scale due to the exponential ILP problem size. Furthermore, for a given deployment scenario, the cost of each possible CPI combination for each query and each attribute that is relevant to that query would have to be determined by the user. We argue that this is not practical in reality. However, the measurements which we showed in Section 3.5.6 indicate that – at least for the measured scenario – inter-dependencies between the used CPIs are weak/non-existent and Securus produces the optimal solution even if the ILP does not correctly reflect the original problem. Thus, if the assumption of independent CPI query latency overheads does not hold in a deployment scenario, Securus may not provide "optimal" solutions in the strict sense, but is able to provide *optimized* solutions.

Policy Expressiveness

Expressiveness of QPs: As shown in Table 3.5, Securus already supports a wide range of SQL syntax elements. The support for additional query components can be added to Securus by analyzing which CPIs are suitable to support them. In order to require that the necessary CPI's ciphertext representation are stored by the CSP that evaluates the query, additional constraints have to be added to the ILP problem similar to those that enforce the already supported query components (cf. Appendix A.2). Thus, Securus constitutes a framework that can be easily extended to support further database query requirements.

Expressiveness of CCs: To provide an impression of the expressiveness of CCs, we show how realistic confidentiality requirements can be expressed via CCs in the following:

- A single attribute is considered confidential. For instance, an attribute *password* that contains passwords of user accounts must not be readable by an attacker. This can be easily expressed by the CC `[password]`.
- Mappings between attributes are considered confidential. For instance, it should not be possible for an attacker to map salaries on names. This requirement can be expressed by the CC `[name,salary]`.
- The anonymity of individual records has to be preserved. For instance, it should not be possible to determine whether the person identified by the *name* "John" or the person identified by "Jack" has a certain *illness*. This kind of anonymity can be guaranteed by specifying the CCs `[name,illness]`. As the CCs guarantees that no CSP can view *name* and *illness*, neither "Jack" nor "John" can be mapped to certain illness. We introduce database anonymization approaches that enforce anonymity without applying CPIs in Section 3.6 and discuss how symbiotic effects between confidential and anonymized database outsourcing can be leveraged by combining both approaches in Section 3.7.
- Functional dependencies between attributes must not be exploitable by an attacker to reveal other confidential attributes or attribute combinations. For instance, if attribute a must not be mappable on attribute b and an attacker is able to reveal plaintext values of attribute a based on values of attribute c, to protect the attribute combination `[a,b]` it does not suffice to protect attribute a. To avoid that an attacker exploits a dependency between attribute a and

c, both attributes have to be considered equivalent. Thus, for each CC that contains attribute a, an additional CC has to be specified that contains attribute c instead of a.

For more examples on how to express confidentiality requirements based on CCs, we refer to Appendix A.4 where we provide a policy profile for the commonly used TPC-C database benchmark [Tra10].

Future extensions to the policy framework: Besides QP and CC requirements, further requirements may exist. In the following we exemplarily list additional requirements and show how Securus can be adapted to support them.

- *Data protection policies*: In some cases, it may not be considered secure to outsource indistinguishable ciphertexts to certain CSPs. For instance, this can be due to the fear that cryptographic schemes will be broken in the future by resourceful attackers such as intelligence agencies that can force the CSP to reveal the outsourced ciphertexts. Thus, a data protection requirement can be: "The attribute X must not be stored outside the European Union." Securus can be extended to support data protection policies in the policy profile by requiring the user to annotate each CSP with its location. The data protection policies can then be satisfied by including constraints in the ILP to enforce that the attributes the data protection policy refers to may only be stored by CSPs that comply to the location requirement of the policy.

- *Trust policies*: In some cases, the available CSPs are not equally trusted with regard to the level of confidentiality that can be expected when outsourcing data to them. For instance, a CSP S_1 may be trusted to handle attribute A in plaintext while other CSPs are assumed to undermine the confidentiality of A. The according trust policy could be: "Attribute A does not require protection at CSP S_1". Securus can leverage such trust policies to increase efficiency. To support the exemplary trust policy, the ILP constraints to enforce CCs at S_1 that contain A can be omitted.

- *Hard efficiency requirements:* In some cases, specific queries have to be guaranteed to be executed within a specific amount of time. Furthermore, an upper limit for the induced transmission overhead may exist. Such hard efficiency requirements can be enforced by limiting the part of the target function that constitute the query latency or transmission cost for a given query to a fixed value (see Section 3.5.4). This can be done by including an appropriate constraint in the ILP problem.

The examples show that Securus constitutes a framework that can be easily extended to support further requirements.

Usability

Specification of relations and QPs: Even without using Securus, the user of a relational database management system has to specify the relations and the indexes that should be created to speed up query execution. Securus makes the same assumption as traditional relational database management systems, namely that the user is able to specify both the data's structure and the QPs of queries that will be executed.

Specification of CCs: Securus aims to provide a mediator that is as efficient as possible and satisfies the user's confidentiality requirements. Thus, Securus requires the user to be able to explicitly state the confidentiality requirements. We argue that three different user types exist:

- Users who have no confidentiality requirements that have to be strictly enforced. Such users can use adaptive onion encryption to opportunistically encrypt attribute values whenever possible (e.g., see CryptDB in Section 3.5.6).
- Users who have confidentiality requirements, but are unable to specify them. If the confidentiality requirements are not known, users have to protect everything and accept the induced efficiency overheads or not outsource data at all.
- Users who have confidentiality requirements and are able to specify them. Based on Securus, such users can leverage their knowledge to outsource data as efficiently as possible without violating any of the confidentiality requirements.

Users that have confidentiality requirements but are unable to specify them can be aided to do so. For instance, Grofig et al. suggest "[...] to create domain-specific templates, e.g. for the financial or healthcare industry, which will help the customer in optimizing his configurations." in [GHH+14]. Thus, an interesting challenge for future work is to investigate how policy profile templates can be generated for specific types of data.

Furthermore, in future work, policy refinement techniques to simplify the specification of CCs can be investigated. For instance, to facilitate the specification of anonymity requirements, the user could specify identifying attributes and sensitive attributes rather than CCs. If a person can be identified by the attributes (ZIP, name) and the attributes (illness, salary) are considered sensitive, these anonymization requirements can be automatically refined to the CCs `[ZIP, illness]`, `[ZIP, salary]`, `[name, illness]`, and `[name, salary]`.

Specification of ICs: ICs express that the attackers do not have the ability to exploit different ciphertext representations of a given attribute. There can be two reasons for this missing ability: 1) the attacker has no background knowledge that can be applied to exploit the outsourced ciphertexts and 2) the potential background knowledge of the attacker cannot be applied based on the outsourced ciphertexts.

Specifying ICs because of the potential attackers' missing background knowledge is not easy in practice. However, based on existing techniques for risk management such as attack trees [MLY05, Sch11] or domain expert opinions, it can be possible to roughly estimate the probability of an attacker having a certain type of background knowledge.

If the user considers the resulting risk acceptable, an IC can be specified. A further exploration of techniques to estimate the risk of attackers having certain background knowledge is subject to future research and beyond the scope of this thesis.

Specifying ICs because background knowledge cannot be applied is easier in many cases. For instance, if an attribute's values are unique, distinguishable ciphertexts of the attribute's values are unique as well. Thus, background knowledge such as the frequency distribution of attribute values cannot be applied to the distinguishable ciphertexts of the attribute and a DIC can be specified for this attribute. In practice, unique attributes occur often. Examples for such attributes include primary and secondary keys in data models. For the data model of the TPC-C benchmark, many ICs can be specified without knowing exactly the semantics of the data as we show in Appendix A.4.

Changing policy profiles after mediator generation: Unlike traditional database management systems, Securus requires that the QPs are specified initially when the database is created. If QPs change, the mediator has to be modified to reflect the changed QPs. Removing QPs that are no longer necessary can be easily supported by deleting the ciphertext attributes that are only needed to evaluate the QP's queries. To add a QP or CC, creating a new mediator from scratch and re-outsourcing the data can be considered. However, this can lead to CC violations as an attacker can have access to both the outsourced data representation of the old and the new mediator. For instance, it is possible that the old mediator encrypted attribute A to satisfy the CC `[A,B]` and the new mediator encrypts attribute B to do so. Thus, an attacker can potentially link the old dataset with the new one and reveal attribute combinations of A and B.

Securus can address this problem by solving a new ILP that contains the constraints of the old ILP along with the additional constraints that originate from the added QP. Furthermore, the allocation of free variables in the ILP solution that has been used to build the old mediator is frozen as a upper bound, i.e., the variable values of the solution have to be put as constants in the new ILP[11]. For instance, given that the solution of the old ILP required the mediator to store distinguishable ciphertexts of attribute A at CSP 1. The solution of the new ILP must not require to store indistinguishable ciphertexts for attribute A at CSP 1, as CSP 1 already knows distinguishable ciphertexts for this attribute.

If a solution to the new ILP problem can be found, the resulting new mediator is backward compatible to the old mediator, i.e., applications can use it just like the old mediator. Furthermore, it efficiently supports queries that are specified by added QPs. If no solution can be found for the new ILP problem, it is not possible to generate a backward compatible mediator without violating CCs. One option to cope with this situation is to apply policy conflict resolution as proposed in Section 3.5.5 and remove conditions from the QPs until the ILP problem is feasible.

[11] Actually, Securus only freezes the variables that are relevant for the solution, i.e., variable allocations that resulted in the utilization of CPIs or the outsourcing of plaintext values to a CSP.

Policy Transformation Performance

Since solving ILP problems is considered NP-hard [Pap81], it is important to evaluate whether the ILP problems that Securus uses to find an efficiency-optimized CPI combination that satisfies the policy profile can be solved in acceptable time. To evaluate that, we generated random policy profiles with different numbers of attributes, QPs, CCs, ICs, and SPs. For each configuration, we generated 100 feasible policy profiles and measured the time needed to find solutions to them. We used the Gurobi[12] ILP solver on a desktop machine with a 2.93 GHz dual core processor and 4GB RAM for our experiments. To get a more realistic impression on the time that is required to find a feasible solution to Securus ILP problems, we measured only policy profiles that are feasible and discarded generated infeasible policy profiles. The time needed by the ILP solver to determine that a policy profile is infeasible was much smaller than the time needed to find efficiency-optimized solutions to feasible policy profiles. Including infeasible policy profiles would have skewed the results and given the false impression that solving ILPs is not time intensive at all.

The results of our measurements are shown in Table 3.6. They show that even big policy profiles with 100 attributes, 80 CCs, 80 QPs, and 6 ICs can be solved in 2560.26 ms on average for 3 CSPs. The maximum observed solving time amounts to 3220.0 ms. As the policy profile only has to be transformed to a mediator once, the results show that Securus' policy transformation is efficient enough to run on consumer hardware.

Notice that ILP problems are considered fixed-parameter tractable with respect to the number of free variables, i.e., if the number of variables is fixed, an ILP can be solved in polynomial time [Len83]. As the full formulation of the ILP problem shows (see Appendix A.2), the number of free variables depends only on the number of attributes in the policy profile, the number of QPs and the number of SPs. Thus, adding CCs and ICs does not exponentially increase the time needed to find a solution for the policy profile.

We also measured the time that is needed to identify policy conflicts in infeasible policy profiles. To do that we generated 100 infeasible policy profiles for each configuration. We avoided generating policy profiles that are trivially infeasible, i.e., policy profiles that contained CCs that conflicted with a single QP. The results of our measurements are shown in Table A.2. They show that it is more time consuming to find a set of conflicting policies in an infeasible policy profile than finding a solution for a feasible policy profile. We measured an average execution time of 174327 ms (~ 3 minutes) for big policy profiles. The longest time Securus needed to find the cause for the infeasibility of a policy profile amounted to 741129 ms (~ 12.5 minutes). We show in Appendix A.3 that policy conflicts can be isolated even more efficiently by relying on Satisfiability Modulo Theory [DMB11] solvers. Based on these results, we consider Securus as efficient enough to detect conflicts on consumer hardware like our utilized desktop computer.

[12] https://www.gurobi.com [last visited on March 2015]

Number of policies/elements					Duration (ms)	
Attr.	QPs	CCs	ICs	SPs	mean	max
10	10	5	1	2	30.39	119.0
10	10	5	1	3	36.64	302.0
10	20	15	2	2	35.30	72.0
10	20	15	2	3	64.24	188.0
40	20	10	2	2	123.59	160.0
40	20	10	2	3	179.48	363.0
50	40	40	3	2	333.67	507.0
50	40	40	3	3	513.89	634.0
60	40	40	3	2	437.24	756.0
60	40	40	3	3	615.67	781.0
80	80	60	6	2	1534.27	2533.0
80	80	60	6	3	2087.49	2703.0
100	80	80	6	2	1871.33	3113.0
100	80	80	6	3	2560.26	3220.0

Table 3.6: Time required to generate a mediator from policy profiles of various sizes.

In future work, the reduction of finding solutions to and policy conflicts in a policy profile on an ILP problem can be further improved. One way to achieve a higher efficiency for finding solutions is to omit free variables in the ILP problem for which the allocation is determined even before the ILP problem is solved. This can also lead to omitting constraints for which the allocation of all contained free variables is already determined.

> *Example:* Consider the case that a CC `[A]` is specified and no ICs are specified on `[A]`. Thus, the free variables $P_{'A',j}$, $O_{'A',j}$, and $D_{'A',j}$ can be omitted and replaced with a constant 0 since they are bound to take the value 0 for $\forall j \in CSPs$ (cf. the full ILP problem formulation in Appendix A.2).

Furthermore, improvements can be achieved by preprocessing the policy profile before reducing it to an ILP problem to split the problem into multiple smaller, independent problems and solve them in a divide and conquer fashion. Thus, each problem can be reduced on its own ILP problem and the final solutions can be combined without jeopardizing feasibility or optimality.

> *Example:* The CC C_1 `[A,B]` and the QP Q_1 `SELECT ... WHERE A=? AND C=?` are independent from the CC C_2 `[X,Y]` and the QP Q_2 `SELECT ... WHERE X=? AND Y=?` if no other CCs and QPs exist. Thus, the attributes A and B along with Q_1 and C_1 can be contained in a ILP problem that is separate from the ILP problem that contains the rest of the attributes, CCs, and QPs.

Number of policies/elements					Duration (ms)	
Attr.	QPs	CCs	ICs	SPs	mean	max
10	10	5	1	2	512.96	976.0
10	10	5	1	3	6713.54	17097.0
10	20	15	2	2	1169.07	1951.0
10	20	15	2	3	11916.88	48937.0
40	20	10	2	2	4440.63	8331.0
50	40	40	3	2	15270.70	51784.0
50	40	40	3	3	36882.95	148477.0
60	40	40	3	2	16842.17	46343.0
60	40	40	3	3	41733.52	179535.0
80	80	60	6	2	66774.91	217159.0
80	80	60	6	3	134166.21	688480.0
100	80	80	6	2	98155.98	327199.0
100	80	80	6	3	174326.99	741129.0

Table 3.7: Time required to find policy conflicts in unsatisfiable policy profiles of various sizes.

3.5.8 Conclusions

In the DaaS problem setting, CPIs can be used to enforce confidentiality requirements and allow the CSP to participate in query execution to increase efficiency at the same time. We surveyed existing CPIs in Section 3.4 and showed against which attackers and for which query requirements they can be applied. In particular, the results of the conducted CPI study show that protecting every information in the outsourced database and providing protection against every attacker is not feasible in most deployment scenarios due to a high efficiency overhead of the according CPIs. Thus, a trade-off between confidentiality and efficiency exists that has to be addressed by applying CPIs selectively. Choosing a set of CPIs by hand, so that the requirements of a given deployment scenarios are met, is possible. However, it requires expert knowledge on each individual CPI and is a complex error prone process, especially when considering databases with many attributes and confidentiality requirements.

Securus constitutes a tunable approach that addresses this problem by allowing to specify the deployment scenario requirements on a high level. Securus automatically determines a set of CPIs that satisfies the requirements. The resulting set of CPIs is implemented in a mediator that can be transparently used to outsource and query data without having to consider the implemented CPIs. Securus maximizes efficiency with regard to query latency and transmission cost by 1) applying the most efficient CPI combination that satisfies the requirements and 2) avoiding the application of CPIs whenever possible. We showed that the optimization problem of finding an efficiency-optimal set of CPIs that satisfy the specified requirements is NP-hard. Securus formalized the optimization problem as an ILP problem and uses established solvers to determine an optimal ILP solution. Based on this ILP solution, Securus makes an efficiency-optimal CPI choice that satisfies the specified confidentiality and query requirements.

In conclusion, Securus constitutes an integrative framework to automate the matching of confidentiality and query requirements with security mechanisms like CPIs. To our knowledge, Securus constitutes the first extensible framework that can provide strict confidentiality guarantees, uses CPIs to satisfy query requirements while not requiring the user to have cryptographic expert knowledge on CPIs, leverages non-colluding storage providers for increased efficiency if available, and aids the user in detecting and resolving conflicting requirements.

The original version of Securus was published in [JKH12]. Compared to [Jü15], we provided an NP-hardness proof for the problem of determining combination of CPIs that satisfy the deployment scenario requirements. Furthermore, we extended the attacker modelling capabilities via inference constraints based on the insights that we gained in the CPI study and extended Securus to support a larger set of the SQL syntax components. Unlike [Jü15] our enhanced version of Securus does not require the assumption that attackers are not able to monitor queries, but only that they are not able to apply background knowledge on query patterns. We extended the ILP problem formulation with the capability to optimize the mediator with regard to query latency and transmission costs. We provided a performance comparison to other confidential database outsourcing approaches and provided guidelines on which approach should be applied in which deployment scenario. Moreover, we added the support to isolate policy conflicts to simplify their resolution.

In future work Securus can be extended to support the requirement to provide protection against attackers with background knowledge on specific queries. Furthermore, Securus can be extended to support anonymity requirements and anonymity enforcement mechanisms additionally to confidentiality requirements and CPIs. We provide a first step in this direction in Section 3.6.

3.6 Anonymized Database-as-a-Service: Dividat

In many deployment scenarios, the anonymity of individuals that are contained in a database has to be protected. Prevalent examples include scenarios in which personally identifiable information (PII) is outsourced, such as hospitals that outsource patient data. It is possible to satisfy anonymity requirements by enforcing data confidentiality via encryption. However, as we showed in Section 3.5, the set of applied CPIs has to be carefully chosen to satisfy the confidentiality and query efficiency requirements. It is not always possible to find a combination of CPIs that satisfies both the query efficiency and the confidentiality requirements. For instance, to our knowledge, no CPI exists to date that allows the CSP to check whether an attribute A's values match a complex regular expression. Thus, based on CPIs it is not possible to protect A and enable an efficient query execution by allowing the CSP to evaluate the regular expression at the same time.

Another way to protect the anonymity of the individuals is to anonymize the database instead of applying CPIs. Unlike encrypted data, anonymized data contains plaintext records on which arbitrary queries can be evaluated. However, all anonymization techniques add noise to the data. This results in larger indexing structures and in the transmission of false-positive records that do not match the executed query. These overheads can outweigh the overhead of transferring encrypted data to the mediator, decrypting it, and evaluating the query at the mediator instead of at the CSP. Thus, the following question arises with regard to optimizing query efficiency and satisfying anonymity requirements at the same time: When should a query condition be evaluated by the CSP based on anonymized data and when can it be considered more efficient to transfer encrypted records to the mediator and evaluate the condition there?

In this chapter, we address this question by proposing Dividat, a concept to build efficiency models for anonymized database outsourcing. We address the problem of index optimization for anonymized databases by making the following contributions:

- We show on the example of ℓ-diversified databases that making query latency predictions for anonymized indexes is hard and optimizing the execution of queries on anonymized indexes can considerably improve efficiency with regard to query latency. In particular, this implies that it is not trivial to choose which query conditions should be evaluated at the CSP instead of the mediator.
- We propose Dividat, **a generic approach to develop models that predict the efficiency costs** with regard to query latency, transmission, and storage. The approach is generic in the sense that it can be applied for all anonymity notions and anonymization techniques. Based on the efficiency models that are developed based on the approach it is possible to optimize the indexing of anonymized data by being able to decide a-priori which indexes are the most beneficial and should be created for a given workload or used to evaluate a given query.
- We provide an **exemplarily utilization of the framework** to derive efficiency models for the case of ℓ-diversified databases.
- We provide a **validation** of the ℓ-diversity specific models as well as the generic framework.

This section is organized as follows. In Section 3.6.1 we provide a generic architecture that can be used to build anonymized database outsourcing approaches. In Section 3.6.2 we show how this architecture can be used to implement ℓ-diversified DaaS and investigate the efficiency of indexes on the ℓ-diversified data. In Section 3.6.3 we propose Dividat, a generic approach to assess the efficiency of anonymized indexes that is independent from the applied anonymity notion and anonymization technique. In Section 3.6.4 we show how the framework can be used to determine efficiency models for the case of ℓ-diversified databases. We show in Section 3.6.5 how the Dividat efficiency models can be applied to optimize the index creation and utilization. The Dividat framework and the ℓ-diversity specific models are validated in Section 3.6.6. Conclusions are presented in Section 3.6.7.

3.6.1 Dividat: Architecture

An overview of how anonymity can be preserved when outsourcing databases is shown in Figure 3.16. Like in the confidential DaaS setting, the user relies on a mediator that is operated by a party that is trusted to enforce the anonymity of individuals in the database. The mediator governs the data outsourcing and the execution of queries. To outsource data records, the mediator encrypts them and stores the encrypted records at the CSP (a). Furthermore, the records are anonymized and stored in index tables (b). The index tables contain anonymized plaintext records. Thus, based on the anonymized index tables, the CSP is able to determine which encrypted records match a query and can send them back to the mediator. The transmitted records may contain false-positive records, i.e., records that do not match the query and occur due to the noise that was added to anonymize the indexes. Thus, the mediator has to "de-anonymize" each record by decrypting it and verify that it indeed matches the query.

3.6.2 Example: ℓ-Diversified Database Outsourcing

In this section we present how ℓ-diversified DaaS can be implemented based on the generic architecture that we introduced in Section 3.6.1 and investigate the efficiency of ℓ-diversified indexes with regard to query latency. We show that the query latency depends on various factors and it is not trivial to predict whether evaluating a query condition on an ℓ-diversified index table improves efficiency compared to transmitting all records to the mediator and evaluating the condition there.

Database ℓ-Diversification

We define that a database is considered ℓ-diverse if each individual QID allocation that is contained in the database maps to ℓ distinct sensitive attribute value combinations, including the true sensitive attribute value combination of the individual[13]. For instance, consider a hospital with a patient *A* who suffers from headache and a patient *B* who has the flu. The patient records in a hospital's patient database are considered 2-diverse if it is impossible to derive from the database whether person *A* or *B* has the illness "Flu" or "Headache".

To ℓ-diversify records, we make use of existing approaches that govern database inserts, updates, and deletions [NC11, NCM13, DFJ+13a]. The approaches provide the CSP with an anatomized view on the data that is ℓ-diverse. The general concept of anatomization is illustrated in Figure 3.17. The figure shows the plaintext database (Figure 3.17a) and the tables that are outsourced to the CSP, i.e., the encrypted records and the anatomized ℓ-diversified data (Figure 3.17b). The attributes that are only contained in the QID of the ℓ-diversity requirements (*Name* in Figure 3.17) and the attributes that are sensitive but not contained in the QID (*Illness* in Figure 3.17) are stored in separate tables. Note that the encrypted records of the main table are stored in the same table as the attributes that are only contained in the QID. Furthermore, each sensitive attribute that is contained in the QID (*ZIP* in Figure 3.17) is stored in a separate table. The attribute values that are contained in the table are loosely

[13] For a more specific definition see Section 3.3.

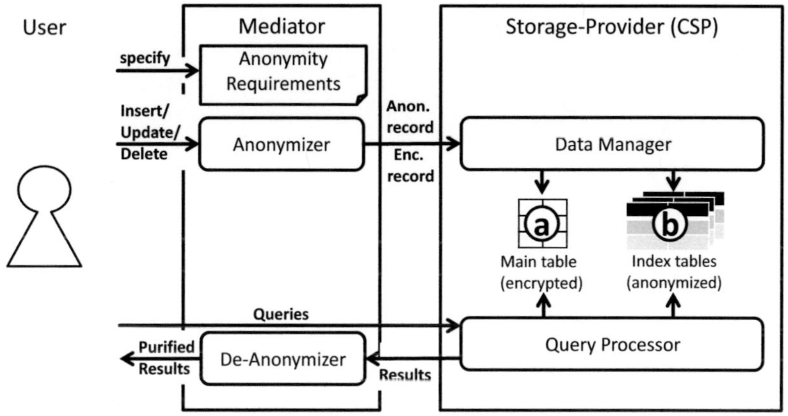

Figure 3.16: Dividat architecture.

coupled via *group identifiers* (GIDs). Each GID links together the distinct attribute values of at least ℓ records of the plaintext database. That way, each attribute value that can serve as a QID maps to at least ℓ different values of each sensitive attribute. Thus, the anatomized data is ℓ-diverse.

> *Example:* In Figure 3.17b, the attribute combination ("John", "12345") can be used as QID and maps to the illnesses "Flu" and "Headache" based in the GID 1. The attribute value "John" for the attribute *Name* that can be used as QID maps to both sensitive *ZIP* attribute values "12345" and "12349" as well as to both sensitive *Illness* attribute values "Flu" and "Headache". As each QID assignment maps to at least 2 different sensitive attribute values, the anatomized dataset is 2-diverse.
>
> To evaluate the query `SELECT * ... WHERE (Name=John OR Name=Dan) AND ZIP=12345`, the mediator rewrites the query to `SELECT Enc ... WHERE (Name=John OR Name=Dan) AND ZIP=12345`. The CSP evaluates this query by using the GID to join the table that contains the attribute *ZIP* with the table that contains the encrypted records and the attribute *Name* (see Figure 3.17b). The CSP evaluates the query on this join and the records *Enc(John, 12345, Flu)* and *Enc(Dan, 23456, Cancer)* are returned to the mediator. The mediator then decrypts both records and discards *(Dan, 23456, Cancer)* as it does not match the original query. Finally, the query result *(John, 12345, Flu)* is returned.

Efficiency of ℓ-Diversified Indexes

Especially in databases with many records, performing joins to evaluate queries is considered expensive with regard to the execution performance of queries. To address this problem we investigate an approach that builds on the existing anatomization solutions and determines in which cases pre-joined anatomized index tables like the one that is shown in Figure 3.17c improve the query execution performance.

To investigate the performance of pre-joined anatomized index tables, we executed queries on index tables that each contained a different set of attributes and measured the latencies of the queries. Surprisingly, an index table that contains all attributes that are relevant to evaluate query conditions is not necessarily the most efficient choice and the evaluation specific query conditions at the mediator instead of at the CSP can be beneficial for query latency as we show in the following.

All measurement were conducted based on a PostgreSQL[14] database that is executed on a machine with a 2.50GHz QuadCore CPU and 4GB RAM. The database that was anonymized and outsourced contained 10.000 records with three attributes. The values of the attributes were chosen uniformly at random. The attribute *gender* could take two possible values, the attributes *synth1* and *synth2* could each take 20 possible values. We set the ℓ-diversity requirements to *QID*={*gender, synth1*}, *sensitive attributes*={*synth1,synth2*}.

Name	ZIP	Illness
John	12345	Flu
Eve	12349	Headache
Marc	12345	Asthma
Dan	23456	Cancer

(a) Plaintext data.

GID	Enc	Name
1	Enc(John, 12345, Flu)	John
1	Enc(Eve, 12349, Headache)	Eve
2	Enc(Marc, 12345, Asthma)	Marc
2	Enc(Dan, 23456, Cancer)	Dan

GID	ZIP
1	12345
1	12349
2	12345
2	23456

GID	Illness
1	Flu
1	Headache
2	Asthma
2	Cancer

(b) Anatomized data at CSP (*anatomized view*).

GID	ZIP	Illness
1	12345	Flu
1	12345	Headache
1	12349	Flu
1	12349	Headache
2	12345	Asthma
2	12345	Cancer
2	23456	Asthma
2	23456	Cancer

(c) Pre-joined index table based on *ZIP* and *Illness* at CSP.

Figure 3.17: Anatomization Example. QID: {Name, ZIP}, sensitive attributes: {ZIP, Illness} [KH14].

14 http://www.postgresql.org [last visited on March 2015]

Definition - Selection [KH14]: A query contains a selection condition on an attribute *a* iff it selects only records that contain one out of a set of values that the attribute *a* can take.

Example: The query `SELECT * ... WHERE ZIP=12345 AND Phone < 35638 AND (Illness=None OR Illness=Flu)` performs selections on *ZIP*, *Phone*, and *Illness*.

For each investigated index table, we measured the latencies of 6000 randomly chosen queries that contained a selection condition on *gender* that selects one value and conditions that select either 1,4,7,10,13, or 16 values for each *synth1* and *synth2*. The average measured latencies are shown in Figure 3.18 with 95% confidence intervals.

The measurements in Figure 3.18a show that evaluating queries on an ℓ-diversified index table that does not contain all relevant query condition attributes can be beneficial compared to index tables that contain all relevant attributes. For instance, evaluating queries that select just 1 attribute value for both *synth1* and *synth2* based on the 5-diversified index table that contains all three relevant attributes *gender*, *synth1*, and *synth2* induces the least efficiency cost (see gender-synth1-synth2 graph in Figure 3.18a). However, if more than 7 attribute values of *synth1* and *synth2* are selected, it is more beneficial to evaluate the query based on an index table that just contains *gender* and *synth1* (see gender-synth1 graph in Figure 3.18a) or even just *gender* (see gender graph in Figure 3.18a).

Furthermore, the measurements in Figure 3.18b show that the required level of diversity ℓ has an influence on when it is more beneficial to evaluate query conditions at the mediator. If 5-diversity is required and 7 values are selected for *synth1* and *synth2*, the gender-synth1-synth2 index table can be considered most efficient. However, if 7-diversity is required, the index table that contains only *gender* is more efficient in case 7 values were selected. Notice that the index table that contains only *gender* is the same for the case of 5-diversity and 7-diversity as it does not contain any sensitive attributes and no joins are necessary. Thus, the measured query latency is identical for both cases.

The conducted measurements indicate that even in simple scenarios the chosen index for a query has a substantial impact on query latency. Furthermore, the measurements show that choosing the optimal index table is not trivial and a variety of factors determine which index table can be considered optimal, including the query itself (e.g., the number of selected values) as well as the anonymization requirements (e.g., the diversity factor ℓ and the QID and the sensitive attributes). To understand the efficiency costs that come with ℓ-diversified DaaS, factors of influence have to be identified and their interactions have to be explored and modeled. We propose a generic approach to develop efficiency models for anonymized DaaS in Section 3.6.3 and exemplarily apply the approach to build efficiency models for the case of ℓ-diversified database outsourcing in Section 3.6.4. Understanding and modeling the interaction between the factors that influence efficiency characteristics such as query latency, storage, and transmission costs is vital to optimize the utilization of anonymized indexes in general and ℓ-diversified indexes in particular. We show how our efficiency models can be used to optimize the creation and utilization of indexes on anonymized data in Section 3.6.5.

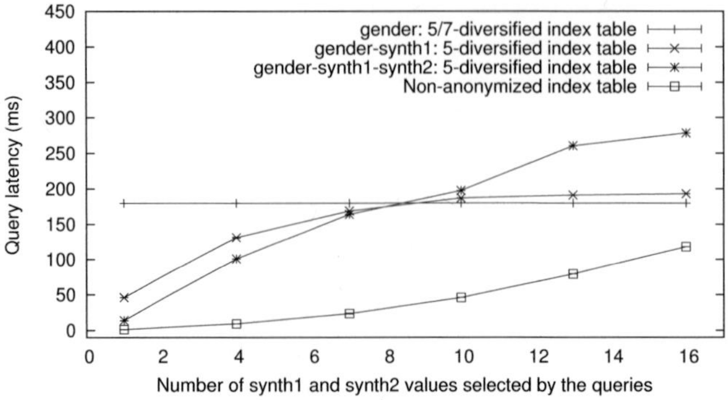

(a) Query latency vs. contained attributes.

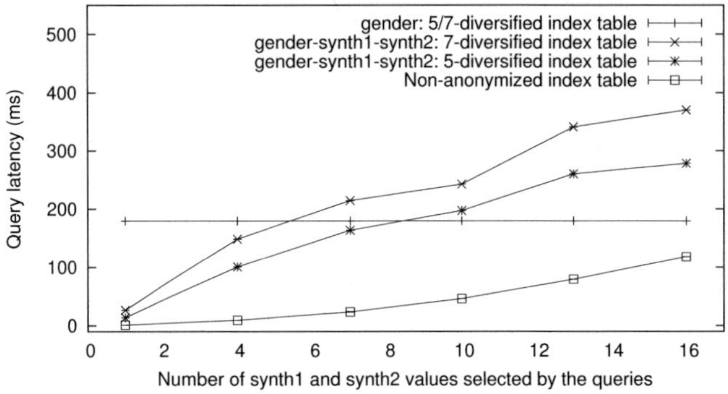

(b) Query latency vs. diversity factor ℓ.

Figure 3.18: Performance differences between indexes with varying parameters.

3.6.3 Dividat: Generic Efficiency Model Framework

Efficiency overheads that are incurred by database anonymization come with respect to several metrics. Dividat focusses on the incurred overheads with regard to query latency, required storage, and data transmission between the mediator and the CSP. An overview of the metrics and the mathematical models that we focus on is shown in Figure 3.19. The storage overhead that is induced by anonymization only depends on the size of the additional indexes that are created based on the utilized anonymization approach and the encryption that is applied to the records in the main table. The transmission overhead that is induced when executing a query depends on the number of records that are transmitted from the CSP to the mediator. The query latency depends on a variety of factors such as database indexing, the I/O operations necessary to determine the records that match a query, network bandwidth, and computation performed by the mediator to decrypt the transmitted records. Due to the systems' complexity, modeling all these influencing factors is not feasible in practical deployment scenarios.

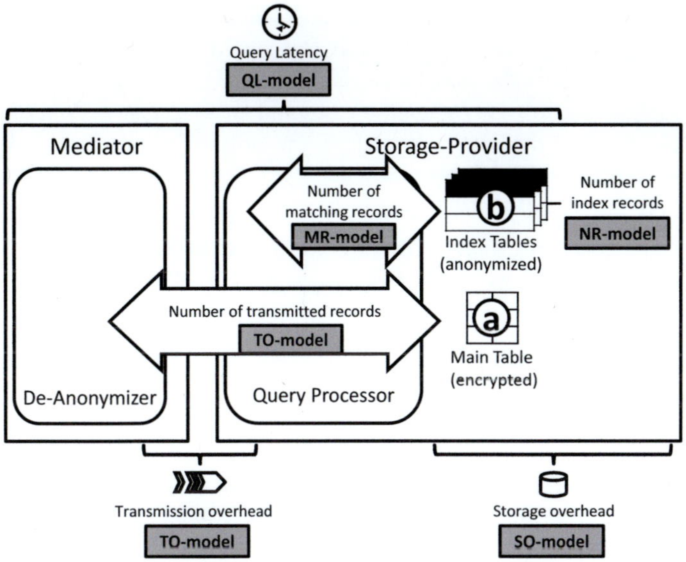

Figure 3.19: Influencing factors for query latency, transmission and storage cost.

In this section we show how it is possible to abstract from the mentioned factors by focusing on the number of records that match a query in the used anonymized index table and the number of transmitted records. We propose models to predict the query latency (QL-model), storage cost (SO-model), and transmission cost (TR-model) that is induced when using anonymized index tables to evaluate queries. The models constitute a generic framework that can be applied regardless of how the database is anonymized and which anonymity notion is achieved. They constitute a framework in the sense that they rely on further models that are specific for the type of anonymity notion and the anonymization technique that is used to achieve this notion. These specific models capture the number of records that are contained in an index table (NR-model), the number of *anonymized records that match* a given query in a given index table (MR-model) and the number of *encrypted records that are transmitted* to the mediator as query result (TR-model). We will exemplarily propose such models for the case of anatomized ℓ-diverse databases in Section 3.6.4.

It may seem that the number of matching records and the number of transmitted records are equal at first sight. However, this is not necessarily the case, as anonymized indexes contain noise[15]. This noise can lead to a situation in which multiple entries in the anonymized index point to the same encrypted record. If a query matches multiple index entries that point to the same encrypted record, the encrypted record is transmitted to the mediator once as the CSP avoids transmitting the same encrypted records multiple times.

15 This is true for ℓ-diversified indexes in particular, but applies to other anonymity notions as well.

Example: In the pre-joined index table shown in Figure 3.17c, the query `SELECT ... WHERE ZIP < 12350 AND (Illness = Flu OR ILLNESS = Headache)` matches 4 records. However, only the 2 encrypted records *Enc(John, 12345, Flu)* and *Enc(Eve, 12349, Headache)* that have the same GID as the matching records in the pre-joined index table are transmitted to the mediator.

The query `SELECT ... WHERE ZIP = 12345 AND Illness = Flu` matches only a single record in the pre-joined index table shown in Figure 3.17c. However, the 2 encrypted records with GID 1 have to be transmitted to the mediator.

The notation we use throughout this section to formalize the proposed models is shown in Figure 3.8 for reference.

Storage Cost Model (SO-model)

By multiplying the number of index table records N_I with the storage space s_I that is needed to store the attribute values of a single record for all attributes $a \in I$ that are contained in index table I, the storage cost that is induced by an index table I can be exactly calculated based on the following analytical model:

$$S(I) = s_I \cdot N_I \tag{3.11}$$

The number of index table records N_I can be predicted by the NR-model which is specific for each anonymization approach.

Transmission Cost Model (TO-model)

The transmission cost, i.e., the number of bytes transmitted from the CSP to the mediator to execute a query depends on the number $t(I,q)$ of records that are transmitted when evaluating a query q on index table I and the size a_q of those records. The transmission cost for evaluating a query q on index table I can be calculated based on the following analytical model:

$$T(I,q) = a_q \cdot t(I,q) \tag{3.12}$$

The size a_q of the transmitted records depends on the outsourced data and thus has to be determined for each query q of the deployment scenario at hand. The number of transmitted records $t(I,q)$ can be predicted by the TR-model which is specific for each anonymization approach. In particular, the overhead compared to evaluating the query on non-anonymized data originates from transmitting false-positive records, i.e., records that do not match the original query but that cannot be filtered out by the CSP based on the anonymized data.

Query Latency Model (QL-model)

The query latency, i.e., the time that is needed evaluate a query from the mediator's perspective, depends on many factors like the used database system, the disk latency and the hardware configuration. To provide a mathematical model that can be applied in practice to determine the latency that has to be expected when evaluating a query on a given index table without modeling each aspect of modern IT landscapes, we abstract from most of these factors by making the following assumption:

N_M	Number of outsourced records
N_I	Number of records in index table I
$S(I)$	Storage cost for storing index table I
$T(I,q)$	Transmission cost for executing query q on index table I
$L(I,q)$	Query latency cost for executing query q on index table I
$r(I,q)$	Number of records that match query q in index table I
$t(I,q)$	Number of records that are transmitted when executing query q on index table I

(a) Notation of the generic models.

$QidOnly$	Set of attributes that are only contained in the QID
$QidSens$	Set of attributes that are both sensitive and contained in the QID
$SensOnly$	Set of attributes that are sensitive but not contained in the QID
A_I	Set of attributes that are contained in the index table I
$R(I,q)$	Set of attributes that are contained in the index table I and that are relevant to evaluate the conditions of query q
q_a	Set of attribute values that match the conditions of query q on attribute a
V_a	Set of attribute values attribute a can have
$p_a(v)$	Relative frequency of attribute value v within all outsourced attribute values of attribute a

(b) Notation of the ℓ diversity specific models.

Table 3.8: Notation for the Dividat models.

Assumption I: *The query latency $L(I,q)$ that can be expected when executing a query q on index table I depends linearly on the number of records $r(I,q)$ that are contained in the index table I and match query q as well as on the number of records $t(I,q)$ that are transmitted to the mediator:*

$$L(I,q) = b \cdot t(I,q) + c \cdot r(I,q) + f \tag{3.13}$$

□

In Section 3.6.6, we provide evidence that Assumption I holds.

To apply the QL-model that is shown in Equation 3.13, the parameters b, c, and f as well as $r(I,q)$ and $t(I,q)$ have to be determined. The number of matching and transmitted records – $r(I,q)$ and $t(I,q)$ – can be predicted by the MR- and TR-model which are specific for each anonymization approach. Modeling b, c, and f analytically is hard, as they depend on a variety of factors (e.g., indexing structures, disk latency, caching effects, database implementation, etc.). We propose to determine them empirically. To determine the parameters for an existing index table I, the measured execution times $e_1, e_2, \ldots, e_n$ of multiple queries $q_1, q_2, \ldots, q_n$ that match and return a different number of records $r(I,q_1) \neq r(I,q_2)$ and $t(I,q_1) \neq t(I,q_2)$ can be measured. Simple linear regression can then be used to determine parameters b, c, and f that provide the best fit for the following system of equations: $e_i = b \cdot t(I,q_i) + c \cdot r(I,q_i) + f, i = 1 \ldots n$. The query latency predictions of the parameterized QL-model are accurate based on assumption I.

3.6.4 Example: Efficiency Models for ℓ-Diversified Databases

In Section 3.6.2, the empirical examination of the efficiency overheads that are induced by ℓ-diversifying databases showed that it is hard to make assertions on which ℓ-diversified index table can be considered efficiency-optimal for which query based on measurements alone. In order to understand the costs and trade-offs that come with anatomization-based ℓ-diversified database outsourcing, an empirical examination as conducted in Section 3.6.2 is not sufficient. An analytical investigation of the efficiency overheads of ℓ-diversified index tables based efficiency models is required to better understand the trade-off between efficiency and ℓ-diversity requirements and to determine an efficiency-optimal utilization of anatomization-based ℓ-diversified index tables. In the following we apply the proposed efficiency model framework to create efficiency models for databases that were ℓ-diversified based on anatomization. To apply the framework, we propose models that predict the number of records contained in a given pre-joined index table (NR-model), the number of records that match a query in an index table (MR-model), and the number of records that are transmitted when a query is executed based on a given index table (TR-model). We show that the models provide very good approximations to results that were measured in real systems in Section 3.6.6.

Number of Index Table Records Model (NR-model)

For a pre-joined index table I that contains the set of attributes A_I, the number of records N_I that are contained in the index table can be calculated based on the following analytical model:

$$N_I = \ell^{|QidSens \cap A_I| + \delta(|SensOnly \cap A_I|)} \cdot N_M \tag{3.14}$$

where $\delta(x) = \begin{cases} 1 \text{ if } x > 0 \\ 0 \text{ else} \end{cases}$ and N_M is the number of outsourced records.

Index table I can be built by joining all tables of the anatomized view on the GID that contain attributes included in A_I with the table that contains the N_M encrypted records. For each sensitive attribute a that is contained in the QID and in the index table ($a \in QidSens \cap A_I$), the table in the anatomized view that contains attribute a has to be joined based on the GID. Each such table contains ℓ records for each GID and, thus, increases the number of entries in the resulting pre-joined index table by the factor ℓ. Furthermore, if a sensitive attribute that is not part of the QID is contained in A_I, i.e., the set $SensOnly \cap A_I$ is not empty, the table that contains the sensitive attributes has to be joined as well. As this table contains ℓ records for each GID, joining the table increases the number of records that are contained in the pre-joined index table by factor ℓ. The rationale behind the NR-model is illustrated in the following example.

> *Example:* An exemplary anatomized view and a pre-joined index table is shown in Figure 3.20 to illustrate the NR-model. An index table that just contains *Name* and no sensitive attributes has the same size as the original database (see the table that contains *Name* in Figure 3.20a).
>
> The pre-joined index table that is shown in Figure 3.20b contains the attributes $A_I = \{Name, ZIP, Illness\}$. To create the index table, all tables of the anatomized view that is shown in Figure 3.20a have to be joined. When joining the table that contains *ZIP* to the table that contains *Name*, the size of the resulting table is $8 = 2^1 \cdot 4 = \ell^{|QidSens \cap A_I|} \cdot N_M$. When additionally joining the table that contains *Illness* with the result, the size of the resulting table amounts to $16 = 2^2 \cdot 4 = 2^{|QidSens \cap A_I| + \delta(|SensOnly \cap A_I|)} \cdot N_M$.

Matching Records Model (MR-model)

Parts of the paragraph of the MR-model are quoted verbatim from [KH14]. The query latency that is induced when evaluating a query q on a pre-joined index table I depends on the number of records $r(I, q)$ that are contained in the index table and match the query. In the following we provide analytical models that can be used by the QL-model to a-priori estimate the number of matching records $r(I, q)$ of a query q that is evaluated based on a given ℓ-diversified index table I. For simplicity, we make the following assumption.

GID	Enc	Name
1	Enc(John, 91231, Flu)	John
1	Enc(Eve, 12349, Headache)	Eve
2	Enc(Marc, 12345, Asthma)	Marc
2	Enc(Dan, 23456, Cancer)	Dan

GID	ZIP
1	91231
1	12349
2	12345
2	23456

GID	Illness
1	Flu
1	Headache
2	Asthma
2	Cancer

(a) Anatomized data at CSP (*anatomized view*).

GID	Enc	Name	ZIP	Illness
1	Enc(John, 91231, Flu)	John	91231	Flu
1	Enc(John, 91231, Flu)	John	91231	Headache
1	Enc(John, 91231, Flu)	John	12349	Flu
1	Enc(John, 91231, Flu)	John	12349	Headache
1	Enc(Eve, 12349, Headache)	Eve	91231	Flu
1	Enc(Eve, 12349, Headache)	Eve	91231	Headache
1	Enc(Eve, 12349, Headache)	Eve	12349	Flu
1	Enc(Eve, 12349, Headache)	Eve	12349	Headache
2	Enc(Marc, 12345, Asthma)	Marc	12345	Asthma
2	Enc(Marc, 12345, Asthma)	Marc	12345	Cancer
2	Enc(Marc, 12345, Asthma)	Marc	23456	Asthma
2	Enc(Marc, 12345, Asthma)	Marc	23456	Cancer
2	Enc(Dan, 23456, Cancer)	Dan	12345	Asthma
2	Enc(Dan, 23456, Cancer)	Dan	12345	Cancer
2	Enc(Dan, 23456, Cancer)	Dan	23456	Asthma
2	Enc(Dan, 23456, Cancer)	Dan	23456	Cancer

(b) Pre-joined index table based on *Name*, *ZIP*, *Illness* at CSP.

Figure 3.20: Example to illustrate the NR-, MR-, and TR-models for ℓ-diversified database outsourcing. The ℓ-diversity requirements are set to: $QidOnly = \{Name\}$, $QidSens = \{ZIP\}$, $SensOnly = \{Illness\}$, $\ell = 2$.

Assumption II: *The attributes' values are independently distributed, i.e., the probability that a database record contains a value* a *for attribute* A *is independent from the probability that the record contains a value* b *for attribute* B. *We discuss this assumption in Section 3.6.6.*

For simplicity and without loss of generality we presume that queries have a specific form that every query can be mapped on. For each attribute that is relevant for a query condition, the query condition that contain the attribute are concatenated with `OR` operators. Conditions on distinct attributes are concatenated with `AND` operators.

Uniform MR-model: Given a database in which all attribute values are uniformly and independently distributed and a required diversity of ℓ. The expected number of matching records for a query q that is evaluated on a pre-joined index table I can be calculated based on the following mathematical model:

$$r(I,q) = \prod_{a \in R(I,q)} \frac{|q_a|}{|V_a|} \cdot N_I \tag{3.15}$$

where V_a is the set of possible values for attribute a, i.e., $|V_a|$ denotes the cardinality of attribute a. The cardinalities of attributes can be maintained in database statistics which are kept by the CSP. The variable q_a denotes the set of values that match the query condition on attribute a. *Selectivity* is the percentage of records that contain specific attribute values. As we assume for now that attribute a is uniformly distributed, the average selectivity of attribute a's values amounts to $\frac{|q_a|}{|V_a|}$. Furthermore, as the attribute values are assumed to be independently distributed, the combined selectivity of multiple attributes concatenated with `AND` operators can be calculated by multiplying the selectivity of single attributes. Multiplying this combined selectivity with the number of records in the anonymized index table results in the estimated number of records that are selected by a query.

The assumption of uniformly distributed attribute values does not hold for real use-cases. For non-uniformly distributed attribute values the model in Equation 3.15 produces inaccurate predictions. To account for non-uniformly distributed attribute values, the frequency of attribute values can be maintained in statistics that are kept by the CSP. In the following we will denote the occurrence probability of value v for attribute a as $p_a(v)$.

Non-uniform MR-model: The expected number of matching records for a query q that is evaluated on an index table I can be calculated based on the following mathematical model:

$$r(I,q) = \prod_{a \in R(I,q)} \sum_{v \in q_a} p_a(v) \cdot N_I \tag{3.16}$$

The selectivity of a condition amounts to $\sum_{v \in q_a} p_a(v)$ as it suffices for a record to contain any attribute value that matches the condition. This corresponds to the `OR` operator that concatenate conditions on the same attribute. The rationale behind the MR-model is illustrated in the following example.

Example: Let us assume that the number of records that match the query `SELECT * FROM table WHERE ZIP=12345 AND Illness='Asthma'` in the table shown Figure 3.20b has to be predicted. The table contains $N_I = 16$ records. From the anatomized view in Figure 3.20a it can be determined that $p_{ZIP}(12345) = \frac{1}{4}$ and $p_{Illness}('Asthma') = \frac{1}{4}$. Thus, $\frac{1}{4} \cdot N_I = \frac{1}{4} \cdot 16 = 4$ records in the table can be expected to match the condition on the attribute *ZIP*. On these records, the condition on the attribute *Illness* has to match additionally. Due to the assumed independency of the attribute value distributions, in expectation, $p_{Illness}('Asthma') \cdot p_{ZIP}(12345) \cdot N_I = \frac{1}{4} \cdot 4 = 1$ record matches the query.

In fact, the query matches two records in Figure 3.20b. The MR-model calculates the *expected* number of matching records in a given index table. The MR-model compensates that, for instance, the query `SELECT * FROM table WHERE ZIP=23456 AND Illness='Flu'` does not match any record in Figure 3.20b.

Transmitted Records Model (TR-model)

The query latency and the transmission cost depend on the number of records $t(I, q)$ that are transmitted to the mediator when evaluating a query q on an index table I. As shown in Section 3.6.3, the number of transmitted records $t(I, q)$ does not necessarily equal the number of index table records $r(I, q)$ that match the query.

The number of transmitted records directly depends on how many distinct GIDs the index table records that match the executed query have. If they have n different GIDs, $\ell \cdot n$ records have to be transmitted, as each GID maps to ℓ distinct encrypted records by definition (cf. Section 3.6.2). In the following we will denote the set of encrypted records that map to a common GID as *GID-block*. To determine the number of transmitted records, the number of GID-blocks (identified by the GID attribute) that match the query has to be determined.

Example: In Figure 3.20b, executing the query `SELECT ... WHERE ZIP=12349 AND Illness=Flu` matches two records that both have the same GID number 1. Thus, the two encrypted records (John and Eve) from the GID-block with the GID number 1 have to be transmitted to the mediator, as the CSP is not able to determine which record that is contained in the block actually has the matching attribute combination. The query `SELECT ... WHERE (ZIP=12349 OR ZIP=91231) AND (Illness=Flu OR Illness=Headache)` matches eight records in Figure 3.20b. However, those eight records all have the same GID number 1. Thus, again only the two encrypted records (John and Eve) that correspond to the GID number 1 have to be transmitted.

The example shows that the number of transmitted records cannot be derived trivially from the number of matching records. In the following we propose a model that can be used to calculate the expected number of transmitted records when evaluating a query q on an index table I. Like the MR-model, the TR-model is based on assumption II, i.e., the assumption that attribute values are distributed independently from each other.

Uniform TR-model: Let's assume that all attribute values are uniformly and independently distributed. The expected number of transmitted records for a query q that is evaluated on an index table I can be calculated based on the following mathematical model:

$$t(I,q) = \prod_{a \in R(I,q) \cap (SensOnly \cup QidSens)} (1 - \prod_{i=0}^{\ell-1} (1 - \frac{|q_a|}{|V_a| - i})) \cdot \frac{N_M}{\ell} \cdot \ell \cdot \prod_{a \in R(I,q)/(SensOnly \cup QidSens)} \frac{|q_a|}{|V_a|} \quad (3.17)$$

A GID-block matches a query condition on a specific attribute a if at least one attribute a value that is contained in the block matches the condition, i.e., at least one attribute a value of the GID-block is contained in the set q_a of selected values for attribute a. Thus, the probability of a matching block amounts to $1 - \prod_{i=0}^{\ell-1}(1 - \frac{|q_a|}{|V_a|-i})$, where $\prod_{i=0}^{\ell-1}(1 - \frac{|q_a|}{|V_a|-i})$ is the probability that no attribute a value within the GID-block matches the query. The probability that the i-th value in the block is not contained in the set q_a of values that are selected by the query amounts to $\frac{|q_a|}{|V_a|-i}$ since an attribute value can only occur once in a GID-block by definition (see Section 3.6.2). As GID-blocks divide all attribute values of the data in groups that contain ℓ values each and every outsourced record contains exactly one attribute value per attribute, the data is divided in $\frac{N_M}{\ell}$ GID-blocks. Each matching GID-block results in the selection of ℓ records. Thus, the expected number of selected encrypted records amounts to $\prod_{a \in R(I,q) \cap (SensOnly \cup QidSens)} (1 - \prod_{i=0}^{\ell-1}(1 - \frac{|q_a|}{|V_a|-i})) \cdot \frac{N_M}{\ell} \cdot \ell$. The CSP can thin out this selection further by evaluating query conditions on attributes that are not considered sensitive and stored along with the encrypted records. As we assume the attribute values are uniformly distributed for now, each such attribute on average selects $\frac{|q_a|}{|V_a|}$ of the encrypted records that are candidates for being transmitted.

> *Example:* Let us assume that the pre-joined index table that is shown in Figure 3.20b is used to evaluate the query `SELECT ... WHERE ZIP=12349 AND Illness=Flu`. As the query contains conditions that select a single value for *ZIP* and *Illness*, $|q_{ZIP}| = 1$ and $|q_{Illness}| = 1$ holds. Furthermore, based on the tables shown in Figure 3.20a it can be derived that $|V_{ZIP}| = |\{91231, 12349, 12345, 23456\}| = 4$ and $|V_{Illness}| = 4$ holds.
>
> A GID-block contains $\ell = 2$ distinct values for each sensitive attribute. Thus, the probability that a given GID-block does not match the condition `ZIP=12349` can be calculated as $(1-\frac{1}{4}) \cdot (1-\frac{1}{3}) = \prod_{i=0}^{\ell-1}(1-\frac{|q_{ZIP}|}{|V_{ZIP}|-i})$ where $(1 - \frac{1}{4})$ is the probability that the first ZIP value in the GID-block does not equal 12349 and $(1 - \frac{1}{3})$ is the probability that the second value also does not equal 12349. Based on this, the number of GID-blocks that match the condition `ZIP=12349` can be expected to be

$(1 - \prod_{i=0}^{\ell-1}(1 - \frac{|q_{ZIP}|}{|V_{ZIP}|-i})) \cdot \frac{N_M}{\ell}$ as $\frac{N_M}{\ell}$ GID-blocks exist by design. Based on our assumption that attribute values are independently distributed from other attribute values, the expected number of GID-blocks that match both conditions can be calculated as $\prod_{a \in \{ZIP, Illness\}}(1 - \prod_{i=0}^{\ell-1}(1 - \frac{|q_a|}{|V_a|-i})) \cdot \frac{N_M}{\ell} = \frac{1}{2}$. Each GID-block contains $\ell = 2$ encrypted records. As no query conditions exist on non-sensitive attributes, $\frac{1}{2} \cdot \ell$ encrypted records can be expected to be transmitted when executing a query like the exemplary one on the index table that is shown in Figure 3.20b.

If the query contains a condition on a non-sensitive attribute, the number of transmitted can be further reduced. For instance, if the query contained additionally the condition `Name='John'`, not all encrypted entries in Figure 3.20b with GID 1 have to be transmitted but only the record that contains the name 'John'. The TR-model considers that by multiplying the expected number of transmitted records based on the evaluation of the conditions on sensitive attributes with the selectivity of the conditions on non-sensitive attributes: $\prod_{a \in R(I,q)/(SensOnly \cup QidSens)} \frac{|q_a|}{|V_a|}$.

Providing a TR-model for the case of non-uniformly distributed attribute values is more complex. As we focus on providing a generic methodology to optimize the indexing of anonymized database outsourcing rather than providing a specialized concept for the case of ℓ-diversity-based database outsourcing, we consider it beyond the scope of this thesis. However, in Section 3.6.6 we evaluate how good the predictions of the proposed model for uniformly distributed attribute values are in the presence of non-uniformly distributed values.

3.6.5 Example: Usage of the Anonymized Index Efficiency Models

Besides providing insights on the trade-off between anonymity requirements and efficiency with regard to query latency, transmission, and storage overhead, the proposed models can also be used to optimize the use of anonymized indexes. In this section we show how efficiency models can be used to optimize anonymized indexing on the example of the efficiency models for ℓ-diversified indexes that were proposed in Section 3.6.4.

The measurements that we presented in Section 3.6.2 and the models that we proposed in Section 3.6.3 and Section 3.6.4 show that the efficiency of ℓ-diverse index tables heavily depends on a) the data's structure, b) the ℓ-diversity requirements, and c) the executed queries. Thus, to optimize the usage of ℓ-diversified indexes, information on these influencing factors is necessary. The ℓ-diversity requirements are well specified and statistics can be kept by the SP to derive the data's structure.

In this section, we show how it can be determined *which index tables should be used to execute a query* and *which index tables should be created.*

Query execution strategy: Which index table is the most efficient choice depends on the query. Thus, for each query q that is executed it has to be determined which index table I can be expected to induce the lowest query latency. This decision has to be made on-demand upon incoming queries. Consequently, the latency estimation has to be performed efficiently and determining the index tables' efficiency by executing the query on each index table is not an option. Based on the proposed models, the query's latency $L(I, q)$ can be estimated for each index table I and the index table with the minimum expected latency can be chosen.

Index creation strategy: The CSP stores records in anatomized tables that allow to create ℓ-diversified index tables on-demand without interacting with the mediator. The set of index tables that should be created depends on the expected future query workload $\mathcal{Q}$.

The storage cost $S(I)$ of a single index table I, the query latency $L(I, q)$, and the transmission cost $T(I, q)$ that is induced when executing a query q on index table I can be estimated based on models that we proposed in Section 3.6.4. The models have to be parameterized with the MR-, TR-, and NR-models which we provided for the case of ℓ-diversified databases in Section 3.6.4. Based on the parameterized models, the query latency and transmission cost that is induced by a given set of index tables $\mathcal{I}$ for a given workload $\mathcal{Q}$ can be calculated as $\sum_{q\in\mathcal{Q}} \min_{I\in\mathcal{I}} L(I, q)$ and $\sum_{q\in\mathcal{Q}} \min_{I\in\mathcal{I}} T(I, q)$, respectively. The storage cost that is induced by a set of index tables $\mathcal{I}$ can be calculated as $\sum_{I\in\mathcal{I}} S(I)$. Query latency, transmission, and storage costs can be jointly considered by applying the weights w_l, w_t, and w_s according to the user's individual preferences:

$$o(\mathcal{I}, \mathcal{Q}) = \frac{1}{|\mathcal{Q}|} \sum_{q\in\mathcal{Q}} \min_{I\in\mathcal{I}}(w_l \cdot L(I, q) + w_t \cdot T(I, q)) + w_s \cdot \sum_{I\in\mathcal{I}} S(I) \tag{3.18}$$

In order to find the optimal set of index tables $\mathcal{I}$ that minimizes the joint cost $o(\mathcal{I}, \mathcal{Q})$, the query workload $\mathcal{Q}$ that will be executed needs to be known. We propose two options to get hold of the workload:

Option 1 - User interaction: The user can specify the queries that need to be efficiently executable. *Precise queries* can be specified that should be executable as efficiently as possible (e.g., `... WHERE 2 < age < 11 AND name = john`). In case the user is not able to specify *precise* queries, it suffices if the user can specify *abstract queries*, i.e., the expected average number of selected values $|q_a|$ for each attribute a for each query (e.g., $|q_{age}| = 8$, $|q_{name}| = 1$).

Option 2 - Dynamic self-optimization: The user might not be able to specify the expected query workload at all. To optimize index creation without any estimates on the query workload, incoming queries can be monitored and the indexing strategy can be adapted to them. For instance, a sliding window of the last x queries can be taken as reference for future queries and be set as query workload $\mathcal{Q}$. Based on $\mathcal{Q}$, a set of index tables that minimizes the cost function in Equation 3.18 can be determined. The CSP can solve the resulting optimization problem in regular intervals to check whether an adaption of the indexing strategy is beneficial[16].

[16] In order to avoid excessive performance loss for inserting, updating, and deleting records in all index tables and to limit the storage cost induced by index tables, the optimization problem should limit the number of index tables to a certain number.

Regardless of whether option 1 or option 2 is chosen to determine the query workload, an optimization problem has to be solved to determine the set of index tables that minimizes the cost function $o(\mathcal{I}, \mathcal{Q})$. The Dividat approach enables to build efficiency models based on which the ℓ-diversified index optimization problem can be grasped. This was not easy before. Solving the optimization problem is beyond the scope of this thesis. However, we provide a formal description of the optimization problem in Appendix B.1.

3.6.6 Evaluation

We argue that the SO- and the TO-model do not require validation as they are analytical, simple, and can be easily parameterized. Given the size of each encrypted record in the database, the cost for transmitting n records can be easily calculated. Furthermore, the storage required to store index tables can be calculated by knowing the number of records in the table and the size of an index table record. We argue that the ℓ-diversity-specific N_I model also does not require further validation as it is analytical and deterministic. Given a diversity factor ℓ, the set of QID attributes and the set of sensitive attributes, the number of records that are contained in an index table can be precisely calculated.

To validate the general QL-model and the ℓ-diversity-specific MR- and TR-model, we conducted a wide range of empirical measurements. All measurement were conducted based on a PostgreSQL[17] database run on a machine with a 2.50GHz QuadCore CPU and 4GB RAM. The dataset that was anonymized and outsourced contained 10.000 records with three attributes.

Our 2^k-factorial experimental design [Jai91] is shown in Table 3.9. We made multiple choices for each parameter and performed measurements for all possible combination of our parameter choices. We varied the diversity factor ℓ, the set of QID attributes and sensitive attributes, as well as the number of values each attribute can take and investigated both uniformly distributed and non-uniformly attribute values distributions. For each parameter combination we measured the query latency as well as the number of transmitted and matching records of 6000 queries that contained selection conditions on 1, 3, 7, 10, 13, or 16 attribute values (marked in gray in Table 3.9).

Validation of the MR-model

To validate the MR-model that we proposed for anatomization-based ℓ-diversified database outsourcing, we compared the number of index table records that matched each executed query of our experimental design with the predictions of the MR-model. The results of this comparison are shown in Figure 3.21. As the plot shows, the proposed MR-model predicts accurately the number of matching records for uniformly and non-uniformly distributed data.

The attribute values of the generated data were chosen independently and, therefore, matched with assumption II. We discuss this assumption later in this section.

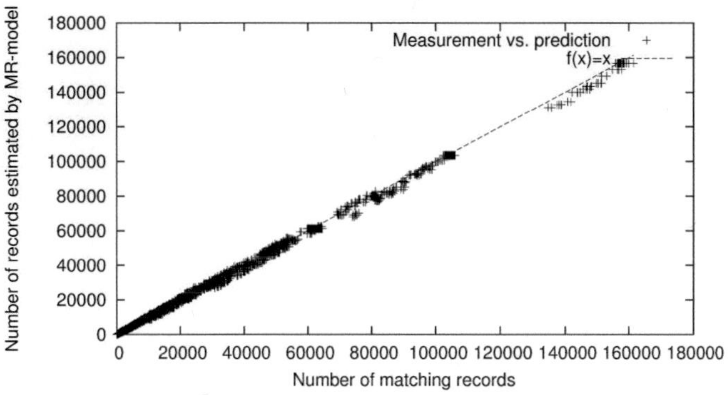

Figure 3.21: Number of matching records vs. MR-model prediction.

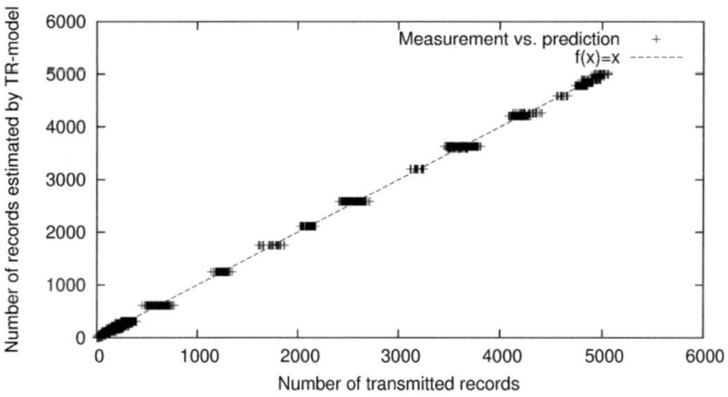

(a) Uniform value distribution.

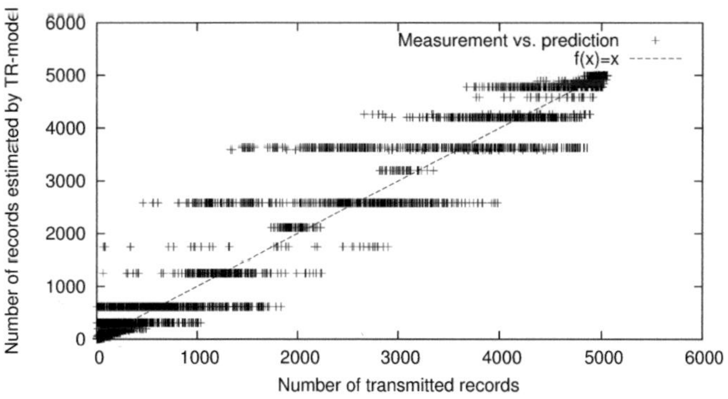

(b) Non-uniform value distribution.

Figure 3.22: Number of transmitted records vs. TR-model prediction.

Anonymity Requirements		Data Structure		Number of selected values					
ℓ	QID sensitive Attr.	Val. Domain Size	Uniform Data Dist.						
ℓ=5	$\|SensOnly\|$=0 $\|QidSens\|$=0	20	yes	1	3	7	10	13	16
			no	1	3	7	10	13	16
		2000	yes	1	3	7	10	13	16
			no	1	3	7	10	13	16
	$\|SensOnly\|$=1 $\|QidSens\|$=0	20	yes	1	3	7	10	13	16
			no	1	3	7	10	13	16
		2000	yes	1	3	7	10	13	16
			no	1	3	7	10	13	16
	$\|SensOnly\|$=1 $\|QidSens\|$=1	20	yes	1	3	7	10	13	16
			no	1	3	7	10	13	16
		2000	yes	1	3	7	10	13	16
			no	1	3	7	10	13	16
ℓ=7	$\|SensOnly\|$=0 $\|QidSens\|$=0	20	yes	1	3	7	10	13	16
			no	1	3	7	10	13	16
		2000	yes	1	3	7	10	13	16
			no	1	3	7	10	13	16
	$\|SensOnly\|$=1 $\|QidSens\|$=0	20	yes	1	3	7	10	13	16
			no	1	3	7	10	13	16
		2000	yes	1	3	7	10	13	16
			no	1	3	7	10	13	16
	$\|SensOnly\|$=1 $\|QidSens\|$=1	20	yes	1	3	7	10	13	16
			no	1	3	7	10	13	16
		2000	yes	1	3	7	10	13	16
			no	1	3	7	10	13	16

Table 3.9: Experimental design of the Dividat evaluation. The measured configurations are marked gray.

Validation of the TR-model

To validate the TR-model that we proposed for anatomization-based ℓ-diversified database outsourcing, we compared the number of records that had to be transmitted for each executed query of our experimental design with the predictions of the TR-model. The results are shown in Figure 3.22 for uniformly and non-uniformly distributed data. Figure 3.22a indicates that the TR-model accurately predicts the number of transmitted records for queries on uniformly distributed data. Figure 3.22b indicates that the TR-model cannot be used to predict accurately the number of transmitted records for queries on non-uniformly distributed data. This is due to the fact that the analytical TR-model that we proposed is based on the assumption that the attribute values are uniformly distributed.

Validation of the QL model

In this section we present a validation of the QL-model that uses the ℓ-diversity-specific MR-model and TR-model as input. To validate the QL-model based on the measurements we conducted, we parameterized the QL-model as explained in

[17] http://www.postgresql.org [last visited on March 2015]

Section 3.6.3. We used the parameterized QL-model to predict the query latency of the executed queries of our experimental design and compared the predicted latencies with the measured latencies. The results for the queries that were executed on uniformly distributed data are shown in Figure 3.23. The plot in Figure 3.23a shows that, besides some outliers, the QL-model predicted the measured query latencies accurately. To provide more insight on the mass distribution of the measurements and the outliers, Figure 3.23b shows the number of queries for which the QL-model predictions deviated x milliseconds from the measured query latencies. Furthermore, the 95% quantiles of the absolute measured prediction errors that are shown in Figure 3.24c show that the prediction error increases relative to the absolute query latency. Accurate predictions are possible even for small absolute query latency. In particular, for 95% of the queries that induce a latency of less than 50 ms, the prediction error is less than 2 ms.

A comparison of the QL-model predictions and the measured latencies for the queries that were executed on non-uniformly distributed data are shown in Figure 3.24. The plot and the histogram can be interpreted like in the case of uniformly distributed data. Surprisingly, despite the fact that the TR-model is inaccurate for non-uniformly distributed data, the QL model made accurate predictions. This can be explained by the parameterization process of the QL-model that was based on the inaccurate TR-model predictions. The linear factors of the QL-model were determined based on inaccurate numbers of transmitted records $t(I,q)$. Apparently, the determined linear factors compensated the prediction errors of the TR-model. To actually use this model for ℓ-diversified database outsourcing with non-uniformly distributed attribute values in practice, further validation is necessary. As we focus on providing a generic framework to develop efficiency models for anonymized database outsourcing, providing a fully validated model for the case of ℓ-diversified database outsourcing is beyond the scope of this thesis.

Influence of the MR- and TR-model on Query Latency

The QL-model makes the assumption that query latency linearly depends on the number of matching and transmitted records. Our measurements show that this assumption holds in the measured scenarios. Figure 3.21 shows that the MR-model accurately predicts the number of matching records and Figure 3.22a shows that the number of transmitted records is also accurately predicted. As the query latencies are accurately predicted based on the QL-model that has been parameterized with the accurate MR- and TR-model (see Figure 3.23) this shows a linear dependency of the query latency on the number of matching records and the number of transmitted records.

We showed that the QL-model makes accurate predictions on the query latency for ℓ-diversified database outsourcing. Now we show that the numbers of matching and transmitted records are indeed necessary input parameters for the QL-model regardless of the used anonymity notion and anonymization technique. For applications in the real world the question arises whether the number of matching and transmitted records is necessary, or if either the influence of the number of transmitted records

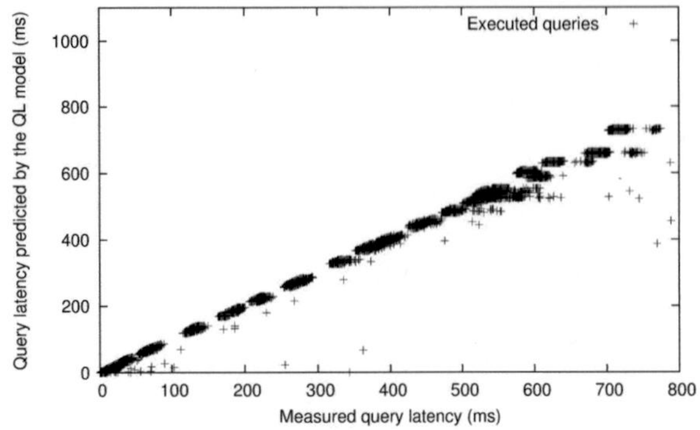

(a) Measured query latency vs. predicted query latency.

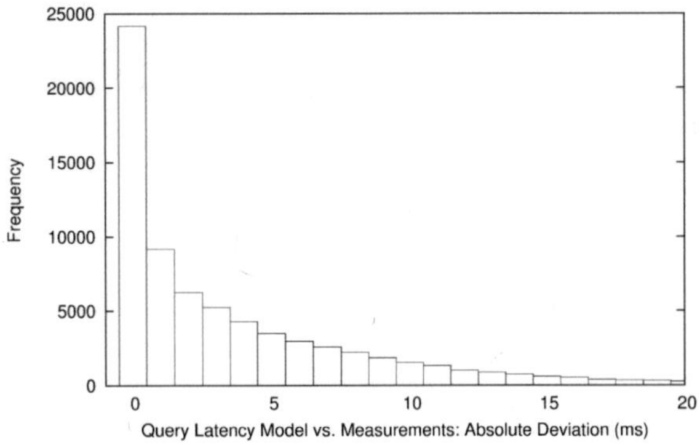

(b) Aggregated deviations: QL-model predictions vs. real measurements on uniformly distributed attribute values (histogram).

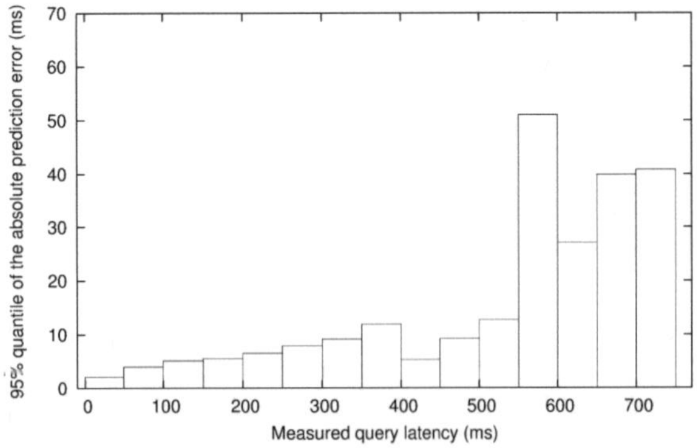

(c) 95% quantiles of the measured absolute prediction error of the QL-model for queries that induce different absolute latencies.

Figure 3.23: QL-model predictions vs. real measurements on uniformly distributed attribute values.

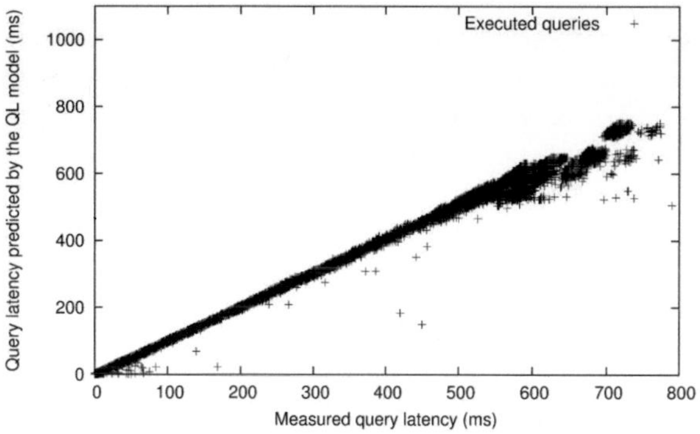

(a) Measured query latency vs. predicted query latency.

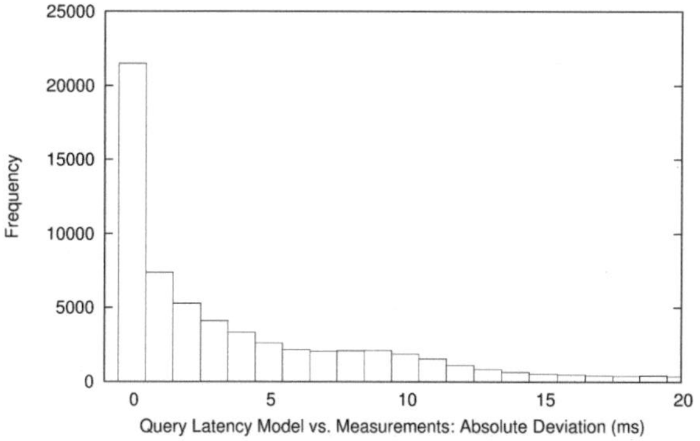

(b) Aggregated deviations: QL-model predictions vs. real measurements on non-uniformly distributed attribute values (histogram).

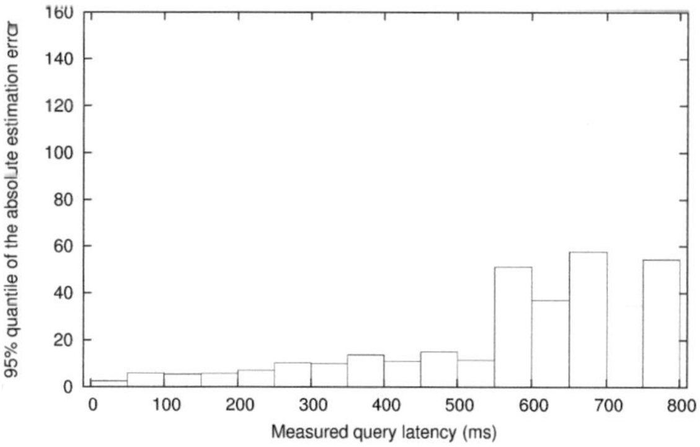

(c) 95% quantiles of the measured absolute prediction error of the QL-model for queries that induce different absolute latencies.

Figure 3.24: QL-model predictions vs. real measurements on non-uniformly distributed attribute values.

or the influence number of matching records dominates so much that it suffices to model one of them in order to build accurate query latency models.

To investigate whether it suffices to make the MR- or the TR-model the only input parameter for the QL-model, we investigated the dependency of the query latency on each, the number of matching and transmitted records. We did that by comparing the number of matching records of each query in our experimental design with the measured query latency. Furthermore, we compared the number of transmitted records with the measured query latency.

The results are shown in Figure 3.25 for the case of a non-anonymized database ($|QidSens| = 0$ and $|SensOnly| = 0$ in Table 3.9) and an anonymized database ($|QidSens| = 1$ and $|SensOnly| = 1$ in Table 3.9).

In case of the non-anonymized database, the measured query latencies depend linearly on both the number of matching records (Figure 3.25a) or the number of transmitted records (Figure 3.25b). In plaintext databases the number of matching records equals the number of transmitted records as no noise is applied and all records that match a query in an index are also transmitted.

In case of the anonymized database, the query latency neither linearly depends on the number of matching records (Figure 3.25c) nor the number of transmitted records (Figure 3.25d). The results show that in general it does not suffice to only take the MR- or the TR-model into consideration and substantial mispredictions are possible. Figure 3.25d shows that the query latency nearly has a linear dependency on the number of transmitted records. Thus, in our particular setup it seems like the number of transmitted records is the dominating factor due to the network and the decryption overhead that is incurred by each transmitted record. However, as Figure 3.25d shows, in case of 5000 transmitted records, not taking the number of matching records into consideration leads to a misprediction of up to 200 ms. This is very dependent on the setup in which the measurements were conducted. With a faster network link and/or a mediator with more computing power (or hardware acceleration for encryption schemes), the network/computation bottleneck will be removed and the MR-model will gain influence. In this case, not taking the MR-model into consideration will lead to substantially worse mispredictions.

Discussion

Functionality of Dividat: We addressed selections of records based on equality and inequality conditions as query components in this paper. However, the proposed concepts and models can be applied to most other SQL selection conditions. For instance, LIKE operators also constitute a condition that matches specific attribute values. The parameters b, c, and f of the QL-model (cf. Section 3.6.3) have to be determined for each type of condition as they might differ depending on indexing approaches that are utilized by the underlying database system used by the CSP.

We considered query latency as one of the main optimization goals in this paper. While this is true for many use cases, applications exist that update the outsourced data at a high frequency. For these applications the performance of INSERT, UPDATE, and DELETE operations should be included in the optimization process. Our QL-

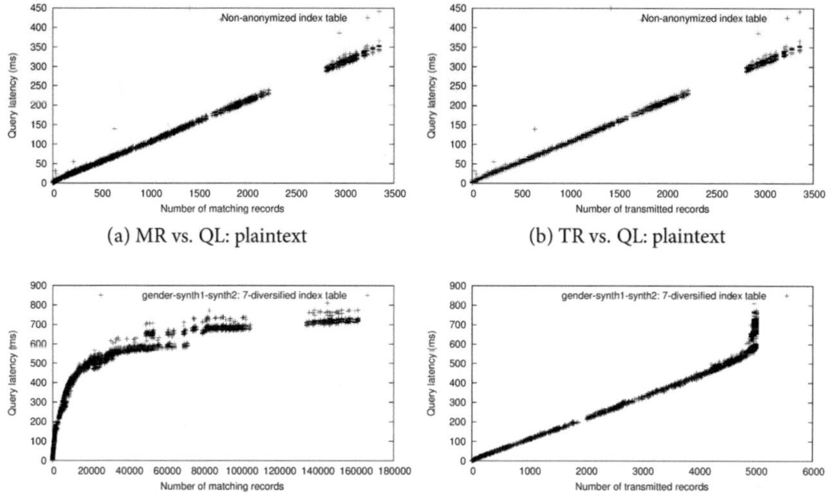

(a) MR vs. QL: plaintext

(b) TR vs. QL: plaintext

(c) MR vs. QL: anonymized

(d) TR vs. QL: anonymized

Figure 3.25: Influence of the number of matching and transmitted records on query latency for plaintext and anonymized databases.

model can be applied to those operations analogously to SELECT operations. The cost function (cf. Equation 3.18) to determine the cost that a set of index tables induces for a given query workload and the optimization problem to minimize the function can be extended to also consider the cost for these additional operations.

Assumption of attribute independency: The MR- and TR-models that we proposed for the case of ℓ-diversified database outsourcing are based on the assumption that the outsourced attribute values are distributed independently from each other. In some deployment scenarios this assumption does not hold. If the attribute value distributions depend on each other, using the proposed models results in a cost prediction error that depends on the strength of the attribute dependency. Even in the field of query optimization the assumption of independent attributes is often made and errors are accepted [Cha98, HILM09]. Thus, this issue remains a challenging problem for future work for query optimization in general. For the applicability of the generic QL-, TR-, and SO-model, the assumption is not necessary.

Overheads induced by applying efficiency models: In order to use the proposed efficiency models to determine for each given query the index table that can be expected to incur the least query latency cost, the CSP has to maintain and query statistical data to optimize the execution strategy for each query. These processes induce a performance overhead. The problem of processing statistical information in traditional query optimization closely resembles the problem at hand and can be applied in this context as well [Cha98, Ioa96, HILM09].

Evolving deployment scenario: The choice which index tables should be created has to be made a-priori based on the predicted future data structure and the predicted query workload. These predictions are not necessarily accurate. We showed that the proposed architecture for anonymized database outsourcing allows the CSP to build index tables without help from the mediator based on a set of anonymized database records (e.g., the anatomized view in the case of the approaches we focused on in this chapter). Thus, once the predicted state differs too much from reality, a new prediction can be made and the indexing strategy can be adapted without inducing any overhead for the mediator [KH14].

Furthermore, the IT infrastructure in the deployment scenario may change over time. For instance, the network link between the mediator and the CSP may become faster. This influences the parameters of the QL model that are determined based on measurements. We argue that parameters can be continuously updated based on latency measurements of the incoming queries. Like for the case of evolving query workload, once the parameters differ too much from the originally determined ones, a new indexing strategy can be determined based on the new parameters without involving the mediator.

3.6.7 Conclusions

One of the main problems that come with building tunable approaches for anonymized database outsourcing is to build anonymized indexing structures that incur the least efficiency costs with regard to query latency, transmission, as well as storage costs and satisfy the anonymity requirements at the same time. To address this problem, it has to be possible to rate index structures according to how much they benefit efficiency when evaluating queries. We proposed Dividat, a generic approach to anonymize outsourced databases and to develop efficiency models for anonymized indexes. The models predict the query latency, transmission, and storage costs that are induced by anonymized indexes. We identified the following three relevant factors of influence for these models: the number of records transmitted between the CSP and the mediator for each query, the number of records that match each query in the anonymized indexes, and the number of anonymized records that are contained in the index. We validated the proposed framework and exemplarily applied it to develop efficiency models for the case of anatomized-based ℓ-diversified database outsourcing. The efficiency models reveal the trade-offs between query latency, storage, and transmission cost as well as the anonymity requirements. Furthermore, the efficiency models can be used to optimize the utilization of ℓ-diversified indexes to reduce query latency, storage, and transmission costs and satisfy anonymity requirements at the same time.

For the case of anatomization-based ℓ-diversified database outsourcing, our results show that adding additional attributes to an index decreases the number of records that have to be transmitted to evaluate a query. However, each added attribute increases the size of the index by up to a factor of ℓ. Against intuition, the number of records in the index table that match a query's conditions are not necessarily decreased but can even be increased by adding attributes on which additional conditions of the query can be evaluated to the index table. An increased number of matching records can potentially result in increased query latency. By applying our generic model framework, we identified the following trade-off situation. The more query conditions can be evaluated based on attributes contained in the index table, the lower is the transmission cost as less false-positive records have to be transmitted to the mediator. However, to enable the evaluation of more conditions, the index table also has to contain more attributes which increase the storage space needed by a factor that depends on the ℓ-diversity requirements. With regard to query latency, adding more attributes is not necessarily beneficial as query latency depends on both the number of transmitted and matching records.

To tune the trade-off between query latency, storage, and transmission cost, we propose to optimize the indexing strategy according to the deployment scenario. The user can weight each efficiency characteristic by its importance in the given deployment scenario or specify strict requirements for each property. The models we proposed and validated to capture the dependency of query latency, transmission, and storage cost on the query and anonymity requirements of the deployment scenario allowed us to formalize the arising optimization problem of finding an optimal set of index tables which satisfies the user's requirements. We leave the development of approaches that solve this optimization problem for future work.

The Dividat framework can be applied to other notions of anonymity such as k-anonymity and t-closeness as well. Regardless of the anonymity notion, overheads occur with regard to unnecessarily transmitted records (false-positives) and the size of indexing structures. To understand and assess the cost of an anonymization technique, models for the number of transmitted records (TR-model) and models for indexing overheads (MR-models) have to be developed. These models can be implemented in the generic Dividat framework to generate efficiency models. Analogously to the case of ℓ-diversified indexing, these efficiency models can be used to optimize the utilization of anonymized indexes.

3.7 Confidentiality-Preserving Indexing vs. Database Anonymization

We provided an overview of two paradigms for database outsourcing: confidential and anonymized DaaS. To achieve confidential database outsourcing, CPIs are required that hide the confidential information while still allowing to efficiently evaluate specific queries. To achieve anonymized database outsourcing, CPIs can also be applied. However, also anonymization techniques that can be considered "weaker" in the sense that they give the CSP access to anonymized plaintext records can be applied to satisfy anonymity requirements. Anonymization approaches outsource noisy plaintext data based on which a CSP can evaluate complex queries. CPIs outsource ciphertext values instead of plaintext values in most cases. With regard to efficiency, each approach has advantages and disadvantages over the other.

Advantages of CPIs over anonymization techniques: For specific query requirements, CPIs can be considered more efficient than anonymization techniques. For instance, it is more efficient to query for records that contain a keyed hash value (`SELECT ... WHERE name=1e2f`) and receiving only matching records than to query an anonymized database for the plaintext value (`SELECT ... WHERE name=John`) and transmitting a lot of false-positive records.

Furthermore, CPIs typically allow to modify the outsourced database easily. While approaches for anonymized database outsourcing can support database modifications as well, update operations typically imply a considerable efficiency overhead for future queries [NCM13, DFJ+13a].

Advantages of anonymization techniques over CPIs: In particular, anonymizing databases instead of applying CPIs can have benefits if strong but inefficient CPIs have to be applied to protect against strong attackers. Furthermore, anonymized databases allow the CSP to participate in the evaluation of more query types such as the search for attribute values that match a regular expression. In the following we highlight in which cases anonymization approaches can be more efficient than CPIs depending on the strength of the assumed attacker, i.e., in which representation the attribute values may be stored at the CSP.

- *Indistinguishable ciphertexts*: If attackers can possess arbitrary background knowledge, only indistinguishable ciphertexts may be stored by the CSPs. To evaluate simple equality selections, searchable encryption or encrypted B-trees have to be applied[18]. Both approaches can induce big efficiency overheads. Based on searchable encryption, the CSP is not able to create indexing structures and has to traverse the entire database for each query which, depending on the number of outsourced attribute values, can incur high query latency overheads. Encrypted B-trees require log(n) rounds of communication and stack network latency (see Section 3.4.4). Depending on the deployment scenario (e.g., the I/O- and computational performance of the CSP and the network link between the mediator and the CSP), anonymized indexes are potentially more efficient.
- *Distinguishable ciphertexts*: If the CSP may store distinguishable ciphertexts of an attribute, deterministic indexes (e.g., keyed hashes) can be outsourced to the CSP. The CSP can use deterministic indexes to efficiently evaluate equality selections. However, to evaluate queries that include range or prefix conditions more inefficient CPIs such as encrypted B-trees are required and anonymized indexes can be more efficient.
- *Order-preserving ciphertexts*: Even if the CSP may store order-preserving ciphertexts, using anonymization techniques to satisfy anonymity requirements can be beneficial. Encrypted indexes cannot support arbitrary queries. For instance, to our knowledge, evaluating arbitrary regular expressions on ciphertexts is not possible up to now. Anonymized indexes contain plaintext data and, thus, can be used to enable the CSP to evaluate arbitrary queries instead of downloading and decrypting the entire database to the mediator.

[18] That is the case if the attacker can be assumed not to monitor queries. Otherwise even more inefficient CPIs such as ORAM have to be applied.

Thus, there are cases in which CPIs can be considered more efficient than anonymization techniques to satisfy anonymity requirements. However, there are also cases in which CPIs are less efficient or cannot be applied to satisfy the query requirements. In these cases, database anonymization can provide an efficiency benefit compared to downloading and decrypting the database and evaluating query conditions at the mediator.

We argue that database anonymization and providing confidential DaaS should be considered together to address the database outsourcing problem setting. This way it is possible to leverage the synergies of the two paradigms by applying both anonymization techniques and CPIs to satisfy anonymity requirements. To leverage these synergies the following question needs to be answered: In which cases is it more efficient to rely on anonymization techniques instead of applying CPIs? In particular, the Dividat efficiency models can be considered as a first step to provide an answer to this question. The development of efficiency models for CPIs is matter of future work.

Furthermore, the question arises of how to build a tunable approach that can tune its properties to satisfy anonymity requirements and optimizes efficiency by relying on CPIs in addition to anonymization techniques. We provide an outlook on jointly considering database anonymization and CPIs security mechanisms by showing how ℓ-diversity requirements and anatomized indexes can be integrated with Securus in Appendix A.7.

3.8 Conclusions

One of the major challenges for secure database outsourcing approaches is the loss of efficiency with regard to query latency, transmission, and storage overhead when applying security mechanisms. In this chapter, we showed how tunable database outsourcing approaches can be built on the example of Securus and Dividat. The proposed approaches maximize efficiency as much as possible without violating the confidentiality/anonymity requirements of the deployment scenario. We showed that the more relaxed the security requirements of a deployment scenario are, the more efficient our approaches were. Our contributions can be summarized as follows.

We contributed a CPI taxonomy and used it to categorize a wide range of existing CPIs. We showed that CPIs differ with regard to the supported query functionality, protection guarantees, and efficiency. We provided a methodology on how to assess CPIs with regard to our taxonomy that encompasses functionality, protection guarantees, and efficiency. In particular, the methodology points out the interdependencies between attacker capabilities, attacker background knowledge, and supported query functionality. To our knowledge, we are the first to propose such a methodology. Based on the conducted survey of existing CPIs we are able to provide recommendations on which CPIs should be applied to protect the confidentiality of attribute values against a given attacker model and allow an efficient evaluation of given queries.

We gave an overview of the Securus approach that allows to tune confidentiality for efficiency. Securus constitutes a framework that optimizes efficiency by tuning its confidentiality properties to satisfy deployment scenario specific requirements. To satisfy a set of specified query and confidentiality requirements, Securus makes use of existing CPIs. We extended Securus to maximize efficiency with regard to query latency and transmission cost by applying the most efficient CPI combination that satisfies the requirements and avoiding the application of CPIs whenever possible. Furthermore, we showed the optimization problem of finding an efficiency-optimized set of CPIs that satisfy the specified requirements to be NP-hard. Securus formalizes the optimization problem as an ILP and relies on established ILP solvers to determine a solution. Based on this solution, Securus makes an efficiency-optimized CPI choice that satisfies the specified confidentiality and query requirements. If some of the specified requirements are conflicting, Securus isolates the requirements that are in conflict and presents them to the user to aid in the resolution of the conflict. We provided a performance evaluation of Securus based on the TPC-C benchmark, compared it to other confidential database outsourcing frameworks, and provided guidelines on which approach should be applied in which deployment scenario. The measurements showed that depending on the confidentiality requirements enforcing database confidentiality is feasible with regard to query performance in practice. To our knowledge, Securus constitutes the first extensible framework that can provide strict confidentiality guarantees, uses CPIs to satisfy query requirements while not requiring the user to have cryptographic expert knowledge on CPIs, leverages non-colluding storage providers for increased efficiency if available, and aids the user in detecting and resolving conflicting requirements.

We proposed the Dividat approach that allows to tune anonymity for efficiency. Dividat constitutes a framework that generates efficiency models for anonymized indexes. Based on such efficiency models, the anonymity of outsourced indexes can be tuned to maximize efficiency and satisfy specified anonymity requirements. Dividat maximizes efficiency with regard to the induced query latency, transmission, and storage cost by optimizing the strategies for anonymized index creation and utilization without violating anonymity requirements. Based on a specified query workload and anonymity requirements, Dividat's efficiency models can be used to determine an efficiency-optimized indexing strategy, i.e., a set of indexes that satisfy the anonymity requirements. We successfully applied Dividat to build efficiency models for anatomization-based ℓ-diversified database outsourcing.

Besides these contributions, we identified addressing anonymity and confidentiality requirements together to leverage the individual benefits of anonymization techniques and CPIs as a promising future research direction. Dividat's efficiency models for anonymized indexes constitute a pre-requisite for including anonymity requirements in the Securus framework. This constitutes a first step towards optimizing the efficiency of anonymized database outsourcing by leveraging CPIs in addition to anonymization techniques. Furthermore, it allows to specify and enforce anonymity *and* confidentiality requirements at the same time.

4 Identity Outsourcing

Federated identity management approaches are widely used already. Users access many web-services not via accounts that they registered with the web-service, but via their Facebook account[1]. In the academic world, literature is provided by library services such as Springer[2] that do not require their users to register for accounts but that make an access control decision based on the account of the users at their home organization.

One disadvantage that comes with federated identity management is that the availability of services for users does not only depend on the availability of service providers, but also on the availability of the user's home organization that maintains the user's digital identity. If the digital identity cannot be provided to the service provider, it is not possible to make an access control decision and, in turn, access to the service is denied. Thus, home organizations have to be highly available to offer highly available services to their users. Being highly available is a challenging task especially for small home organizations that do not operate highly available servers and infrastructure. With the advent of cloud computing such availability challenges are increasingly addressed by outsourcing services and data to highly available clouds. However, outsourcing identities to clouds implies a high risk as IT security strongly depends on the integrity of identity information to correctly enforce access control for services, resources, and data. External providers are not always trusted to enforce security characteristics such as the confidentiality and integrity of identity information. In this chapter we explore how security characteristics can be enforced before outsourcing identities to highly available providers. We consider this as an important step towards the vision of being able to securely outsource entire IT service landscapes without having to worry about by whom they are hosted.

1 https://developers.facebook.com/docs/facebook-login [last visited on March 2015]

2 http://link.springer.com/ [last visited on March 2015]

This chapter is organized as follows. First, we provide an overview of the identity federation concept and highlight its benefits and challenges in Section 4.1. In Section 4.2, we show the prevalent requirements that federated identity management approaches have to satisfy in real scenarios. Federated identity management approaches such as the one that we present in this chapter should ideally be able to tune their properties to satisfy individual requirements of each participant in the given identity federation. We show that trade-offs have to be made with regard the properties of an identity federation approach. In particular, the satisfiability of individual requirements is affected by the federation trust model, i.e., the existing trust relationships between participating user home organizations and service providers. We state the main research question that we address in this chapter in Section 4.3 and give an overview of related work in Section 4.4. In Section 4.5 we present Occasio, an approach that allows to outsource identities to an highly available party that is not trusted to enforce data confidentiality and integrity. Occasio allows to tune security properties in order to satisfy requirements with regard to other quality properties that we identified to be relevant in federated identity management deployment scenarios. We summarize our contributions to the domain of federated identity management and draw conclusions in Section 4.6.

Parts of the contributions presented in this chapter have been previously published in [KH13].

4.1 The Concept of Identity Federations

Federated identity management decouples the task of delivering IT services from the task of having to manage digital identities of users that are eligible to access the service. In identity federations, service providers (SPs) rely on digital identities that are managed by identity providers (IdPs). In this thesis we use the terminology for identity management that is introduced in the ISO/IEC 24760 standard [ISO11a].

> *Definition - Identity Information (Digital Identity) [ISO11a]:* Set of attribute values related to an entity.

> *Definition - Identity Federation [ISO11a]:* Agreement between two or more domains specifying how identity information will be exchanged and managed for cross-domain identification purposes.

> *Definition - Identity Provider (Home Organization) [ISO11a]:* Entity that makes available identity information. [For simplicity, we use the term "identity provider" to refer to the software component that provides the service provider with identity information in this thesis. We use the term "home organization" for the organization that issues the identity data that is provided by the identity provider.]

> *Definition - Identity Information Authority [ISO11a]:* Entity related to a particular domain that can make provable statements on the validity and/or correctness of one or more attribute values in an identity.

In particular, the identity information authority can be the home organization of the user. For instance, a home organization that grants group memberships to its users constitutes the identity information authority for group attributes. Furthermore, the user can be the identity information authority for some attributes. For instance, the user is the identity information authority for the attribute eye color or her residential address.

> *Definition - Service Provider (Relying Party) [ISO11a]:* Entity that relies on the verification of identity information for a particular entity.

Consider an identity federation that contains an SP S_1 that trusts two IdPs I_1 and I_2. A user U_1 that has an account at I_1 can access the service offered by SP S_1. In this setting, user U_1 authenticates to IdP I_1. IdP I_1 asserts that the user successfully authenticated and additionally passes identity information on the user to S_1. Based on the identity information and the assertion that the user successfully authenticated, SP S_1 can make an access control decision and grant access to the user U_1. Note that SP S_1 may encounter user U_1 for the first time, but can nevertheless grant access because it trusts IdP I_1 to correctly authenticate users and provide correct user identities. Therefore, SP S_1 is relieved of the task of having to manage digital user identities. Likewise, a user U_2 that has an account at IdP I_2 can also access the service offered by SP S_1 based on her account. Identity federations have benefits which we highlight in Section 4.1.1 but also challenges which we introduce in Section 4.1.2.

4.1.1 Benefits of Identity Federations

Federated identity and access management allows users to access services that are provided by distinct SPs based on a single digital identity that they have established at their home organization. This results in several benefits.

From the user's perspective, only **one authentication credential** is needed to access multiple services as they all rely on the same account at the user's home organization. This increases usability for the user as she only has to manage and memorize a single credential.

Another benefit in terms of usability is the **single-sign-on** functionality that many identity federation approaches provide. Once a user authenticated at her IdP to access a service that is provided by an SP, the services of other SPs within the federation can be used without authenticating again.

SPs can potentially expect a **higher quality of identity information**, as it is easier for users to update the identity data at a single location and they do not have to keep track of multiple accounts. For instance, if a user moved from one place to another, only the user's digital identity at the home organization has to be updated with the new residential address instead of multiple local accounts at each service. This also implies that the SP does not have to take any quality management actions to ensure up-to-date identity information.

A benefit for both the SP and the users is that the **home organization performs the identification** of the user, i.e., if required, the home organization checks whether the user of an account is indeed who she claims to be, for instance, by physically checking the user's passport. Thus, the user does not have to take any additional steps to verify her identity to the SP and the SP does not have to establish workflows for identifying users.

We show in Section 4.2.2 that some of the listed benefits depend on the amount of trust that exists between the SPs and the IdPs of an identity federation.

4.1.2 Challenges of Identity Federations

Relying on federated identity and access management also induces challenges.

Trust is needed between the participants of an identity federation. On the one hand, SPs rely on the IdPs to correctly authenticate users and to make correct assertions on these users both in terms of identity information such as e-mail addresses and authorization tokens such as group memberships. On the other hand, IdPs rely on the SPs to not misuse the released identity information. Establishing the necessary trust relationships is paramount and not easy in many scenarios. In Section 4.2.2 we investigate further which trust relationships can exist in identity federations.

Integration effort that is required to integrate identity federation approaches with the existing IT landscape has to be considered when federating access to IT services. In this regard, both technical and organizational aspects need to be considered. For instance, consider a user that accesses a terminal server via a Secure Shell (SSH) client. Such a client typically sends the user's password to the SSH server to authenticate the user. However, in the federated case, the password should ideally be sent to the IdP instead of the SP. If the client does not support this, the client has to be modified. Furthermore, when first accessing a service of a foreign organization, the user is obliged to give consent to acceptable-use policies by the provider. In the example of the SSH client this can be problematic as such clients do not offer a rich user interface to display the policies and request the user's consent.

The **freshness of authorization tokens** that are managed by the IdP has to be guaranteed. In the traditional intra-organizational setting, the SP manages authorization tokens such as group memberships and can be sure that the authorization tokens are up-to-date. However, if the authorization tokens are managed by the IdP, assuring that each authorization decision of the SP is based on up-to-date authorization tokens is important. This implies that authorization tokens have to be queried from the IdP for each authorization decision. Furthermore, it has to be guaranteed, that no outdated authorization tokens can be injected.

Service availability no longer only depends on the SP but also on the availability of the IdP. Once the IdP is not available, users that want to access the service can no longer authenticate against the IdP and no authorization tokens can be issued for the SP to make an authorization decision. As the SP cannot decide whether to grant access to the user, the service is not available to the user.

4.2 Identity Federation Deployment

To deploy the concept of identity federations in real scenarios, technologies are necessary that allow to delegate the authentication of a user to the IdP and exchange identity information between IdPs and SPs. In particular, the exchange of identity information is required to allow the SPs to make authorization decisions based on the identity information. Building and extending technologies that establish identity federations is not trivial as requirements of different stakeholders have to be considered. These stakeholders include the SPs, the IdPs, and the users of the identity federations. In this chapter, we propose an approach to outsource identity providers to external parties. Like every other federated identity management approach, our approach has to consider prevalent deployment scenario requirements of the stakeholders of an identity federation. Based on our experiences from the bwIDM project (see Section 2.4), we introduce common deployment scenario requirements that have to be considered when building federated identity management approaches in Section 4.2.1. An identity federation approach has to satisfy requirements that depend on the specific deployment scenario at hand. The deployment scenario encompasses the trust relationships between participants of an identity federation, i.e., the federation trust model. The federation trust model has a significant impact on whether specific deployment scenario requirements are satisfiable. We provide an overview of different federation trust models, highlight their implications on the satisfiability of specific deployment scenario requirements, and categorize existing identity federations according to their federation trust model in Section 4.2.2. Identity federation approaches should be tunable to the federation trust model so that other deployment scenario requirements can be satisfied if the federation trust model permits it.

4.2.1 Prevalent Deployment Scenario Requirements

In this section we provide a set of common requirements for federated identity management approaches. As shown in Figure 4.1 the requirements can be interpreted from the perspective of the different stakeholders who participate in an identity federation, including the user, the SPs, and the IdPs. The list of scenario requirements that is presented in the following must not be understood as exhaustive but is meant to provide an impression on the variety of requirements that have to be considered when building federated identity management approaches. Based on our experiences in the bwIDM project (see Section 2.4), the listed requirements are highly relevant in real deployment scenarios.

Performance Efficiency Requirements:

- **Time behavior:** From the perspective of the user, the time required to perform a login is important. Upper limits for the time behavior of the login process can be a requirement of SPs that want to offer their users a good user experience. Furthermore, SPs can require that it has to be possible to exchange identity information with the IdP in a specific amount of time. For instance, if a HPC service executes a workload that frequently needs up-to-date identity information, minimizing the time that is required to exchange identity information between the IdP and the SP is paramount.

- **Resource utilization:** From the IdP's perspective it is important that the digital identities of its users can be efficiently maintained. For instance, updating existing digital identities should not require an excessive amount of computation, network, or storage resources. Furthermore, the amount of resources required to execute a user login should be kept reasonable low. Each stakeholder has such requirements and they strongly depend on the deployment scenario. For instance, if users should be able to access services via inefficient mobile devices, the federated login process has to be designed in such a way that no excessive computational burden is put on the user client.

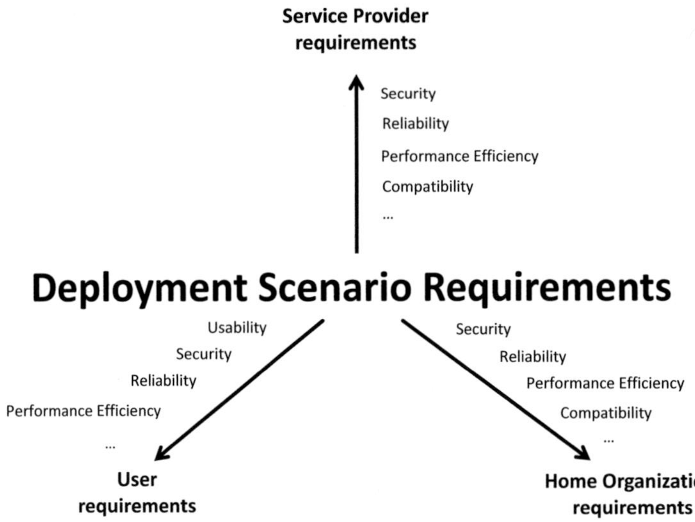

Figure 4.1: Possible deployment scenario requirements that need to be considered when applying federated identity management approaches.

Usability Requirements:

- **Use of home-organizational credentials:** A user that accesses a service based on her home-organizational identity expects to be able to use the credentials that are linked to this identity.

- **Alternative authentication methods:** Some services are extensively accessed via other authentication methods than passwords. For instance, to access terminal servers via SSH, public-private key pairs are often used for authentication. The option to utilize alternative authentication methods should not be eliminated by applying an identity federation approach.

- **Transparency of the approach:** The approach that is used to federate the service should be transparent to the user. Users are accustomed to the traditional service access procedures via local accounts. Ideally, compared to this procedure, the user does not need to perform any additional steps to access the service in a federated way. This includes the utilization of familiar service clients. The established and familiar client software to access the service should be supported by the identity federation approach. For instance, in case of an SSH client, the user should be able to use an off-the-shelf client and not be obliged to download a modified SSH client to access a federated service.

Compatibility Requirements:

- **Interoperability:** From the SP's and IdP's perspective, an important requirement can be that the identity federation approach can be integrated with the IT environment. For instance, if other SPs and IdPs are already operational, a requirement can be that the new identity federation approach should be compatible to them. Another requirement could be that the identity federation approach has to be compatible to established services that formerly relied on other means for authentication and authorization such as LDAP servers (see Section 2.4).

- **Co-existence:** Another requirement can be that the identity federation approach must not have a detrimental impact on specific other technologies that are applied in the deployment scenario. For instance, if an important component that needs identity information runs into frequent timeouts because the identity federation approach cannot provide the identity information fast enough, the identity federation approach and the component cannot co-exist.

Reliability Requirements:

- **Availability:** SPs and IdPs typically aim to provide their users highly available access to services. Thus, requirements can exist with regard to availability of user authentication and identity information. In particular, ensuring availability constitutes a common challenge of identity federations as the service's availability from the perspective of the user does not only depend on the availability of the service provider but also on the availability of the user's IdP. In particular, an identity federation approach can improve service availability by providing the user with options to access services even when her IdP is not available.

Security Requirements:

- **Confidentiality:** The enforcement of the confidentiality of identity information in transit can be considered important by the user to preserve privacy. For instance, the user might trust her IdP and the SPs to view her identity information. However, the confidentiality of the identity information has to be protected when it is transferred from the IdP to the SP.

- **Integrity:** The enforcement of the integrity of identity information in transit can be important for SPs to avoid tampering with authorization tokens and authentication messages. Identity information integrity does not only encompass the correctness of the identity information, i.e., the ability to check whether it was altered, but also the freshness of it, i.e., the ability to check whether the identity information is up-to-date.

- **Authenticity:** Enforcement of user authenticity is important for the SP to avoid unauthorized access to the provided services. Enforcing authenticity often goes hand in hand with the enforcement of integrity. If an IdP correctly authenticates a user, but the message to the SP that asserts a correct authentication can be manipulated, authenticity can be undermined.

All three security requirements are important to the users, the SPs, and the IdPs to preserve the trust relationships within the federation. Even if a user entrusts both an SP and her IdP with her identity information, it would not be in her interest to participate in an identity federation where her identity information can be eavesdropped when it is transmitted from the IdP to the SP. Likewise, an SP might trust an IdP to provide correct identity information on a user but the SP would not rely on an identity federation if the identity information can be manipulated when it is transmitted from the IdP to the SP.

Security requirements have to be satisfied by enforcing specific security characteristics via security mechanisms. Using security mechanisms can conflict with other requirements as we show in the remainder of this chapter. In some cases, enforcing specific security properties by applying security mechanisms is not necessary due to existing trust relationships. We investigate possible trust relationships and their impact on security requirements in the next section.

4.2.2 Identity Federation Trust Model

In this thesis we say "entity *A* trusts entity *B* to perform specific actions" if *A* willingly relies on *B* to perform these actions and accepts the risk of *B* not performing these actions. We use the term "trust" in a black and white fashion [VCDC11], i.e., an agent within a system is either trusted to perform specific actions or she is not. A federation trust model states which participants of a federation trust which other participants with regard to performing specific actions. The federation trust model can have a significant impact on whether specific deployment scenario requirements are satisfiable. In the following we categorize trust relationships that can occur in identity federations and show how the federation trust model of a deployment scenario can affect usability and availability of an identity federation approach. Furthermore, to show that the federation trust models occur in real deployment scenarios, we present a categorization of existing identity federations according to the trust relationships that exist between the participants.

Credential Trust Model

Concerning the user's home-organizational credentials, we differentiate between the case in which the SP may view them (full-trust credential model) and the case in which the SP may not view them (limited-trust credential model).

Full-trust credential model (see Figure 4.2a): In the full-trust credential model, SPs are allowed to view the user's home-organizational credentials as they are trusted to keep them confidential. For instance, such credentials include passwords. A user can send her home-organizational password to the SP to authenticate. The SP can then send the password to the home organization which asserts its validity and, thus, authenticates the user. With regard to usability, the full-trust credential model allows the users to use the passwords of their home-organizational account even with standard (unmodified) service clients that send the password directly to the SP for authentication. For instance, the full-trust credential model can be assumed in case the service is provided by the home organization of the user as, thus, the password does not leave the home organization when authenticating against the service.

> *Example:* In the bwIDM federation, each IdP can decide whether it supports user authentication via the full-trust credential model approach and thus provide a high usability to the user. For instance, for a user to access services that are provided by her home organization, the user's password can be transmitted to the SP as it *is* the user's home organization.

Limited-trust credential model (with home-org. password) (see Figure 4.2b): In the limited-trust credential model, SPs are not allowed to view home-organizational user credentials as they are not trusted to keep them confidential and potentially abuse them. In particular, attackers who compromised the SP must not be able to eavesdrop user credentials. To authenticate via home-organizational credentials, these credentials have to be sent to the home organization first. The home organization then asserts that the user successfully authenticated to the service provider.

While the limited-trust credential model offers higher security guarantees, the user client has to support the process of authenticating against the IdP to access the service of the SP. For web-based services that are accessed via a web-browser this process can be supported by using the browser's redirect capabilities. However, many traditional service clients (such as SSH clients) do not support this process. Thus, users potentially have to use modified clients to access services and usability is affected.

> *Example:* In the DFN AAI federation[3] user credentials may only be send to the home organization of the user. Up to now, the DFN-AAI federation contains only web-based services. Thus, the mentioned usability issues do not occur for DFN-AAI services.

Limited-trust credential model (with service spec. credentials) (see Figure 4.2c): Instead of relying on home-organizational credentials, service-specific credentials can be used to authenticate users with unmodified service clients.

[3] https://www.aai.dfn.de/ [last visited on March 2015]

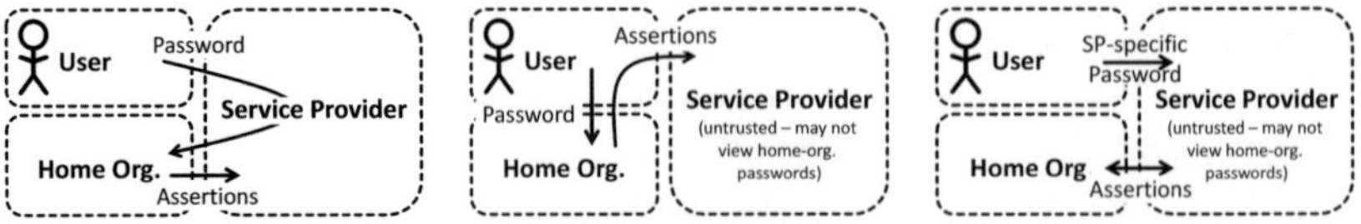

(a) Utilization of the home-organizational password in the full-trust credential model [KSNH13].

(b) Utilization of the home-organizational password in the limited-trust credential model [KSNH13].

(c) Utilization of a service specific password in the limited-trust credential model.

Figure 4.2: Credential trust model.

For instance, these service-specific credentials can be service-specific passwords or public-keys that are deployed at the SP to enable a challenge response authentication. While not using the home-organizational credentials to authenticate with the home-organizational account can be confusing for the user, the users are used to authenticate via service-specific credentials in many cases. For instance, deploying public-keys at terminal servers to access them via challenge response authentication without having to enter password credentials is common practice.

> *Example:* In Grid services such as bwGrid or EGI, user certificates are utilized for user authentication. User's can apply for a user certificate at their home organization. The user certificates are service-specific credentials in the sense that they are only valid for the grid services.

Impact of the credential trust model on usability: Depending on the deployment scenario, the assumed credential trust model can limit the degree of usability that can be achieved by an identity federation approach. If a limited-trust credential model is assumed, common service clients that do not support an authentication against the IdP instead of the SP have to be either modified or the user has to establish and maintain additional credentials with each SP.

Authorization Token Trust Model

To make authorization decisions, SPs need authorization tokens of a user, i.e., attribute values of the user's digital identity such as group memberships. Either the SPs manage these authorization tokens on their own, rely on authorization tokens that are provided by the user's home organization, or rely on authorization tokens that are managed by a third party.

Authorization tokens managed by the SP (see Figure 4.3a): In case the SP manages the authorization tokens that are necessary for the SP's authorization decisions, the SP does not need to trust any external party to correctly maintain the authorization tokens. However, the SP has to create and maintain the authorization tokens on its own. In some cases, this is feasible, in other cases this induces a large overhead for the SP and the user. For instance, if a service requires the users to be affiliated to their home organization as "member" the user has to prove to the SP somehow that she

(a) Authorization tokens managed by the SP.

(b) Authorization tokens managed by the home organization.

(c) Authorization tokens managed by a third party.

Figure 4.3: Authorization token trust model.

is indeed a member at her home organization. Furthermore, if the user's affiliation changes, the SP has to guarantee that the authorization token is updated as well.

In case only members of certain user groups should be able to access a service resource and these groups are locally built at the SP, managing the according authorization tokens is comparatively easy.

> *Example:* Services often use Facebook Connect[4] only for authentication and manage their own authorization tokens. For instance, an SP can authenticate a user based on the user's Facebook account and then authorize the user by checking whether the user already paid to use the service based on a local authorization token "paid for the service until date X". Another example are services that rely on the eduroam[5] RADIUS [RWRS00] federation for authentication. As RADIUS does not support the transmission of authorization tokens, the SP has to manage them on its own.

Authorization tokens managed by the home organization (see Figure 4.3b): In case the SP relies on authorization tokens that are managed and maintained by the home organization, the SP does not need to bear the overhead for managing the authorization tokens. However, the SP needs to trust the home organization to issue correct authorization tokens. In case the service can be misused or damaged by users, it is important for the SP to make sound authorization decisions. If authorization is only performed to limit the amount of utilized resources that the home organization of the users is accounted for, the SP does not need to "believe" in the correctness of the authorization tokens that are issued by the home organization. In case the home organization provides wrong authorization tokens that lead to a wrong authorization decision, the home organization has to pay for the utilized resources.

Note that the concept of relying on authorization tokens that are managed by the home organization can also be used to delegate the whole authorization *decision* to the home organization. For instance, this can be done by defining a single authorization token "may_access_service_X" at the home organization and the authorization rule "if may_access_service_X then grant access" at the SP.

4 https://developers.facebook.com/docs/facebook-login [last visited on March 2015]

5 https://www.eduroam.org/ [last visited on March 2015]

Furthermore, authorization tokens that originate from the home organization can be used in combination with SP-local authorization tokens. Thus, the concepts of token management by the home organization and token management by the SP are not to be understood as mutually exclusive.

> *Example:* In case of bwIDM, the decision which user may access which HPC service is made not by the SP but by the home organizations of the users. For instance, if a home organization issues the authorization token "may_access_bwUniCluster" for a user, that user is granted access to the HPC service bwUniCluster. In case of the DFN-AAI SAML federation, library services make authorization decisions based on the *common-lib-terms* entitlement. If a home organization issues this entitlement for one of its users, this user may download manuscripts from the library service. The download is then accounted to the home organization which pays for the download.

Authorization tokens managed by a third party (see Figure 4.3c): In some cases the authorization tokens that are necessary for authorization decisions cannot be provided by all IdPs within an identity federation or not all IdPs are not trusted to maintain them correctly. If several SPs require these authorization tokens, the overhead of managing authorization tokens redundantly at each SP can be avoided by letting a third party that is trusted by all SPs manage them. After the user is authenticated via its home organization, the necessary authorization tokens for this user are queried from the trusted third party.

> *Example:* Grid environments such as the European Grid Infrastructure (EGI) are comprised of multiple SPs (grid sites) at different organizations. All SPs rely on the same set of authorization tokens. The authorization tokens of a user on which each grid site bases the authorization decision are issued by trusted third parties, so-called Virtual Organizations (VOs).

Impact of the authorization token trust model on availability: "Outsourcing" the management of user authorization tokens by relying on authorization tokens that are issued by the home organization or a trusted third party relieves the SP of the overhead to manage and maintain the authorization tokens. However, the availability of the services provided by the SP depends on the availability of the home organization or the trusted third party. If they are not available, users that try to access the service are denied access as their authorization tokens cannot be retrieved and no authorization decision can be made.

Identity Assurance Trust Model

Federated identity management approaches allow to authenticate users based on their home-organizational digital identity. However, authenticating users based on a digital identity does not guarantee that the user is the natural person that the digital identity claims and that may access the service. The natural person has to be linked to the digital identity, i.e., the digital identity has to be *assured*. For instance, this can be achieved by requiring the user to appear in person and to show a piece of identification before a user account is created.

Home-organizational identity assurance (see Figure 4.4a): If the SP trusts the home organization to correctly identify its users or if a correct user identification is not required, no additional overhead is induced for both the SP and the users. SPs do not have to implement any workflows to assure the user identification and the users can access the services based on their already assured home-organizational accounts without having to assure their identity again.

> *Example:* The bwIDM and the DFN-AAI federation require the home organizations to correctly identify the natural persons before creating digital identities for them. This requirement is manifested within the federation policies that are signed by the participating home organizations. In case of the EGI federation, to get a user certificate that is necessary to access resources, the user has to identify via passport to her home organization. The SPs trust the home organizations to correctly perform this step.

Service provider identity assurance (see Figure 4.4b): In many cases the SP does not trust the home organization to identify its users. However, the home organization is trusted to correctly authenticate digital identities. Thus, to assure the identity of a user, the SP has to create the link between the digital identity and the natural person. For instance, this can be achieved by requiring the users to appear in person and show a passport to the SP or by assuring the identity via other digital identities that are trusted by the SP to belong to the user in question.

> *Example:* Many services use Facebook Connect to authenticate users. If a mapping to a natural person is necessary, additional measures such as showing a piece of identification have to be taken.

Impact of the identity assurance trust model on usability and performance efficiency: If an SP does not trust a home organization to identify its users, the SP has to assure the identity. Methods used to assure identities can constitute a significant usability overhead for users. For instance, if users from the university of Ulm have to show a passport to the SP in the university of Karlsruhe to access its services, users from the university in Ulm will most likely no longer use the service. Furthermore, introducing an organizational workflow to assure identities requires additional resources from the SP. In the example, someone has to be paid to control the passports of users who want to assure their identity.

Categorization of Existing Federations

To show that the introduced federation trust models occur in real deployment scenarios, we provide a categorization of existing federations according to the identified trust models in Figure 4.5. In the bwIDM federation [KLS+14], identity providers can decide

(a) Home-organizational identity assurance.

(b) Service provider identity vetting.

Figure 4.4: Identity assurance trust model.

for themselves whether specific SPs can be trusted to view user credentials (full-trust credential model). Furthermore, SPs rely on authorization tokens that are provided by the users' IdPs and trust in the IdPs to provide correct authorization tokens. In the bwIDM federation, the digital identities of users are assured by requiring the home organizations to identify their users via passport before establishing the digital identity.

The DFN-AAI[6] and edugain[7] federations are bigger than the bwIDM federation and do not assume a full-trust credential model, i.e., user passwords must not be passed to SPs but only to the users' IdP. Thus, users authenticate to their IdP which asserts to the SPs that the user successfully authenticated.

Grid computing federations such as bwGrid[8] and the European Grid Infrastructure (EGI)[9] also assume a limited-trust credential model and require their users to authenticate via challenge response protocols based on issued certificates. Once authenticated, additional identity information and authorization tokens are queried from an attribute provider[10] rather than the home organization of the user.

The eduroam federation[11] is a global federation of universities and primarily used to allow members of a university to access wireless LAN services of other universities. Eduroam can be used only to authenticate users. If an SP requires authorization tokens it has to manage them.

Facebook Connect[12] or Google+ Sign-in[13] allows SPs that offer web-based services to implement a user login based on the user's Facebook or Google accounts. As the identity information provided by Facebook or Google is managed and maintained by the user herself, authorization tokens have to be managed by the SP. Furthermore, the identities provided by Facebook or Google are not assured by, for instance, requiring the users to show a passport when establishing a Facebook account. Thus, if an assured identity is important for an SP, the SP has to perform the identity assurance.

As a rough trend it can be observed that the trust in federation participants decreases with increasing federation size. In small federations such as bwIDM where

6 https://www.aai.dfn.de/ [last visited on March 2015]

7 http://services.geant.net/edugain [last visited on March 2015]

8 http://www.bw-grid.de [last visited on March 2015]

9 http://www.egi.eu/ [last visited on March 2015]

10 One such attribute provider can be the Virtual Organization Membership Service (VOMS) which is prevalently used in grid computing scenarios.

11 https://www.eduroam.org/ [last visited on March 2015]

12 https://developers.facebook.com/docs/facebook-login [last visited on March 2015]

13 https://developers.google.com/+/ [last visited on March 2015]

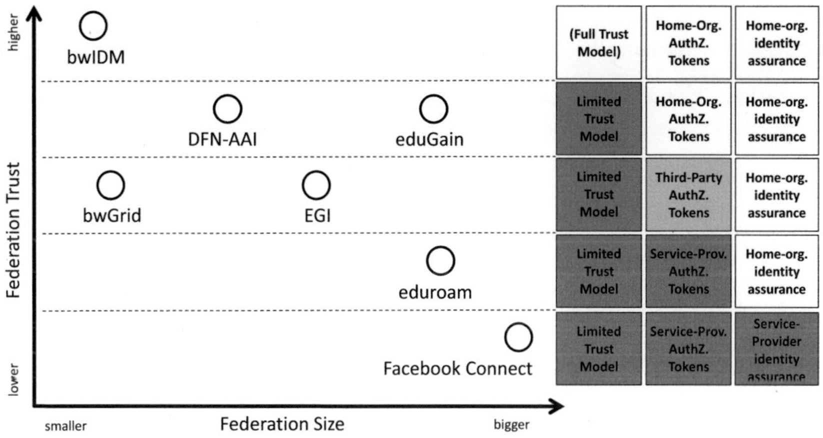

Figure 4.5: Categorization of existing federations according to their federation trust model.

the members know each other well, strong trust assumptions can be made that are not realistic in bigger federations such as Facebook Connect where arbitrary SPs are able to join the federation and the quality of identity information is not guaranteed. This observation underlines the importance of tunable approaches for federated identity management. A static approach that is optimized for small federations where participants have strong trust relationships is not applicable in federations where only weak trust relationships exist. Considering the fact that federations tend to grow, trust relationships may even change over time. It is important that the identity federation approach which constitutes the foundation of the identity federation can be adapted to changing deployment scenario requirements.

4.3 Specific Research Question: Outsourcing Identity Providers

One disadvantage of identity federation is that the availability of a service does not only depend on the availability of the SP, but also on the availability of the user's IdP. A service is unavailable to the users of an unavailable IdP, as the IdP cannot authenticate users and provide the service provider with authorization tokens.

To guarantee high service availability to the users, IdPs have to be operated on highly available platforms. High availability is one of the benefits that specialized cloud providers can offer. This makes the outsourcing of IdPs attractive in case no highly available in-house platform can be provided by the home organization of the users. However, as the outsourced identity provider manages authorization tokens and authenticates users, a malicious cloud provider that operates the outsourced IdP can influence the authorization decisions of each relying service provider. In particular, this implies that an SP that relies on an IdP does not only have to trust the home organization of the user, but also the party that hosts the IdP.

With regards to the problem of ensuring the availability of identity providers, we address the following research question in this chapter:

How can identity providers be securely outsourced to potentially malicious parties that offer highly available infrastructures?

An approach that addresses this question should be tunable to satisfy deployment scenario specific requirements like those that were listed in Section 4.2 and that result from scenario specific federation trust models that were introduced in Section 4.2.2.

4.4 Related Work and Fundamentals

4.4.1 The SAML Standard

We provide the fundamentals of the Security Assertion Markup Language (SAML) standard [HCH+05] for federated identity management. The foundation of a SAML federation constitutes the *SAML Federation Metadata* which is maintained by the administrator of the federation. The SAML Federation Metadata explicitly states which IdPs and SPs are part of the identity federation and how they can communicate with each other. Furthermore, the SAML Federation Metadata contains a public-key for each participant of the identity federation to enable the enforcement of communication confidentiality and integrity between the participants of the federation. The main building blocks of the SAML standard are *SAML-Assertions*, *SAML-Protocols*, *SAML-Bindings*, and *SAML-Profiles*.

SAML-Assertions contain assertions of the IdP on the digital identity of the user. In particular, SAML-Assertions can contain *SAML Attribute Statements*, which contain statements on the user's attribute values, and *SAML Authentication Statements*, which contain the statement that the user successfully authenticated herself against the IdP. SAML allows to enforce confidentiality by encrypting Attribute and Authentication Statements based on the target SP's public-key that is contained in the SAML Federation Metadata of the identity federation. Furthermore, the integrity of Attribute and Authentication Statements can be preserved by letting the IdP sign them based on the IdP's private key. Encryption and signing can also be performed on the level of SAML-Assertions.

SAML-Protocols specify how identity information can be queried by the SP and responded by the IdP. For instance, the SAML Assertion Query and Request protocol can be used by the SP to retrieve SAML-Assertions from the IdP.

SAML-Bindings specify how the SAML-Protocol messages can be transported from the SP to the IdP and vice versa. For instance, the SAML HTTP Redirect Binding can be used to relay the SAML-Protocol messages over the user's web-browser when the user aims to access a web-based service.

SAML-Profiles combine SAML-Assertions, SAML-Protocols, and SAML-Bindings to use-cases. A commonly used SAML-Profile is the SAML-WebSSO profile. The SAML-WebSSO profile allows users to access web-based services based on their home-organizational digital identity. When the user tries to access a service, the SP uses the SAML Assertion Request protocol to request SAML-Assertions from the user's

IdP. The SAML Assertion Request message is passed to the IdP via a SAML HTTP Redirect Binding, i.e., the user is redirected to her IdP's login page with the SAML Assertion Request. Once the user authenticated successfully, the IdP issues SAML-Assertions that contain SAML Attribute Statements on the user's digital identity and a SAML Authentication Statement which states that the user authenticated successfully. These SAML-Assertions are packed into a SAML Response protocol message which is relayed back to the SP via the SAML HTTP Redirect Binding. Once the SP receives the SAML-Assertions, it uses them to check whether the user successfully authenticated and whether the statements on the user's digital identity are sufficient to grant access to the user. If this is the case, the user is granted access to the service.

The contribution that we present in this chapter enforces security properties when outsourcing IdPs to external parties. This is orthogonal to the scope of the SAML standard which focuses on establishing identity federations and standardizes the interaction between IdPs and SPs. We show how our approach can be integrated with the SAML standard in Section 4.5.6.

4.4.2 Outsourcing Identity and Access Management

Leveraging federated identity management in the context of cloud-computing was proposed in many publications [KIT07, NBKT11, SSMJF12]. In particular, the benefits that originate from the decoupling of identity providers and service providers in the cloud computing scenario are stressed. While some of the proposed approaches also to leverage the high availability of cloud providers by outsourcing digital identities, they assume that the providers can be trusted and do not modify or view identity information to manipulate access control decisions. The Occasio approach allows to leverage the benefits of outsourcing identity providers even if the cloud provider is not trusted to correctly enforce the confidentiality and integrity of identity data. It achieves that by enforcing the integrity and confidentiality of identity information before it is outsourced to the untrusted provider.

Entity-centric approaches [ABR+10, RBO+10, RSN12, BPFS09] and attribute based credentials [CDL+13, SKR12] as a special form of entity-centric approaches let users store their own identity information and, thus, do not rely on potentially unavailable identity providers. Besides burdening the user with the obligation to store identity information, updating and revoking authorization tokens based on entity-centric approaches constitutes a hard problem. Attributes that are part of a digital identity can constitute authorization tokens such as group memberships. Such authorization tokens must only be issued by a party that is trusted by the SPs (see authorization token trust model in Section 4.2.2). In most cases, this trusted party is not the user. A user can be prevented easily from manipulating the attributes of her identity by letting the trusted party apply a cryptographic signature. However, if the trusted party has to update an attribute, the user has to be available to receive the update. Furthermore, the old version of the attribute has to be revoked, i.e., flagged as no longer valid. Approaches that address revoked attribute-based credentials exist [CKS10, CL02, LKDDN10], however, they incur big continuous computational overheads on the party that issues the attribute-based credentials.

Occasio has some resemblance to entity-centric approaches in the sense that the user's home organization does not participate in the process of user authentication and authorization. However, compared to existing entity-centric approaches, Occasio relieves the user from having to store her own digital identity and cryptographic secrets[14]. Furthermore, Occasio is capable of providing the guarantee that attributes have not been revoked in the last *t* seconds. Unlike in attribute-based credential revocation approaches [CKS10, CL02, LKDDN10] this time window can be chosen arbitrarily small without inducing much computational overhead on the party that issues and maintains the identity attributes of the users.

Integrity preserving database-as-a-service approaches: The idea to outsource identities to highly available yet untrusted parties can be considered as a special case of database outsourcing. Protecting integrity against malicious storage providers was investigated in the database outsourcing community. Approaches were proposed to enforce completeness and correctness [DGMS01, MNT06, NT06] as well as freshness [XWYM08, XWYM07] of query results. A query result is *complete*, if all records that match the query are contained in the results. The query result is *correct* if no data attributes are manipulated and it is *fresh* if no records are contained in the results that are no longer up-to-date. Signature aggregation and chaining techniques [MNT06, NT06] can be used to enforce correctness and completeness. Enforcing freshness based on those techniques is costly with regard to efficiency as, depending on the number of database records, a large number of signatures have to be updated. The use of Merkle Hash Trees (MHTs) [Mer89, Mer79] has been proposed to enforce freshness efficiently [DGMS01, GSMB03, MNT06]. However, achieving completeness based on MHTs for arbitrary queries and applying frequent updates to the outsourced data is expensive [DGMS01, NT06, XWYM08] which makes the use of MHTs to enforce database integrity unattractive in many database-as-a-service deployment scenarios. When outsourcing identities, relying on MHTs is a good choice to enforce correctness and freshness as completeness does not need to be enforced. This is due to the fact that omitting an outsourced identity and thereby claiming that an identity has never been outsourced cannot be used to grant a user more access rights than she actually has in most deployment scenarios. Furthermore, for the case of identity outsourcing we showed that the update efficiency and throughput that is achieved when relying on MHTs is acceptable.

Wei et al. published an approach that improves the efficiency of storing and retrieving MHT signatures from relational databases [WY14]. Occasio makes use of similar techniques to improve the performance of updating and retrieving outsourced identities. Compared to Occasio, Wei et al.'s approach only focusses on improving the efficiency of MHTs and was published after Occasio.

Occasio maps the problem setting of identity outsourcing to the setting of secure database outsourcing. It shows how security mechanisms that originate from the database outsourcing domain can be used in the federated identity management context and how the security mechanisms can be integrated with existing concepts

[14] This is true if SPs can be trusted to handle passwords correctly (see Section 4.5.8).

and standards of identity federations. Furthermore, Occasio makes use of additional security mechanisms to address user authentication based on the outsourced digital identities which is beyond the scope of traditional database-as-a-service approaches that focus on the preservation of data confidentiality and integrity.

4.5 Securely Outsourcing Identity Providers: Occasio

To address the research question that is stated in Section 4.3, we propose Occasio, a tunable approach that can be used to securely outsource digital identities to third parties that are not trusted to enforce the data confidentiality and integrity. If required, Occasio selectively applies security mechanisms before outsourcing the identity information. Thus, Occasio can be tuned to satisfy different deployment scenario requirements. In particular, this allows Occasio to tune the trade-offs that were identified in Section 4.2.2.

This section is organized as follows. We give an overview of the identity outsourcing problem setting in Section 4.5.1. In particular, we introduce the requirements that an identity outsourcing approach has to satisfy additionally to those provided in Section 4.2.1. In Section 4.5.2 we provide an overview of the Occasio approach. In Section 4.5.3, Section 4.5.4, and Section 4.5.5 we show which security mechanisms can be applied to satisfy the deployment scenario security requirements. In Section 4.5.6 we evaluate how the security mechanisms can be integrated with the SAML standard. We evaluate the efficiency of the security mechanisms in Section 4.5.7. The results of these evaluations show that the security mechanisms induce a trade-off between security characteristics and efficiency, usability, and compatibility. We list those trade-offs explicitly in Section 4.5.8 and show how Occasio can be tuned to satisfy given deployment scenario requirements. We discuss the Occasio approach in Section 4.5.9. Finally, we summarize our findings and conclude the chapter in Section 4.6.

4.5.1 Problem Setting

A conceptual overview of the problem setting of outsourcing identity providers to an external, highly available *identity provider representative* (IdPR) is shown in Figure 4.6. The home organization (HO) outsources statements on the identity information of its users to the IdPR. A statement on a user assert that the user's digital identity contains specific attribute values. Since the SP can retrieve the statements from the IdPR, the HO does no longer have to be available to provide identity information when users access the service.

In the following, we describe the problem setting in more detail and highlight the possible requirements that an identity outsourcing approach has to satisfy. These requirements have to be considered as an addition to the generic requirements that we presented in Section 4.2.1. Which requirements have to be satisfied depends on the deployment scenario. In particular, the deployment scenario includes the assumed federation trust model as well as the assumed attackers.

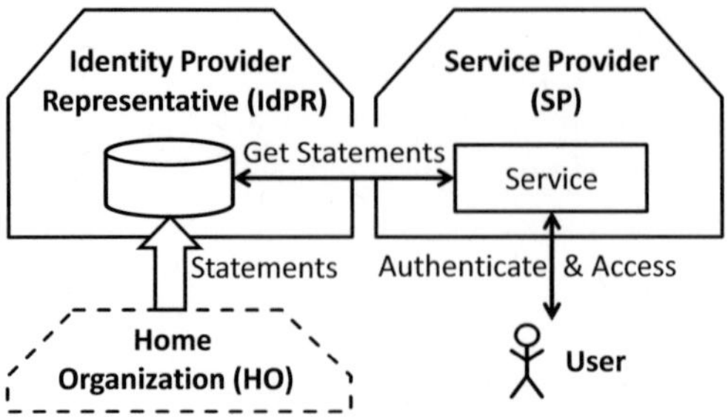

Figure 4.6: Problem setting of identity outsourcing [KH13].

- **Requirement 1 - No HO involvement:** The HO must not be required to be available during user logins. If the HO had to participate to authenticate/authorize a user to use the services of an SP, the SP would not be available once the HO is not available. A high degree of availability can only be guaranteed if the HO is not required to participate in the process of user authentication and of making an authorization decision when a user accesses services.

In the following, we initially assume the following worst case attacker model: The IdPR, the SPs, and the users are assumed to be malicious in the sense that they aim to read outsourced statements without permission of the HO. Furthermore, they try to alter statements or inject statements that have already been revoked by the HO to impersonate users or enable specific users to access services they are not authorized to. To reach their goals, the IdPR, the SPs, and the users are assumed to collaborate. The HO is trusted by the SPs to provide correct and up-to-date statements on its users.

This worst-case attacker model results in the following requirements that have to be satisfied.

- **Requirement 2 - Confidentiality of identity information:** The confidentiality of the outsourced statements has to be enforced in the sense that they are only accessible by the HO and authorized SPs. Not every SP may view every outsourced statement. The HO has to obtain the user's consent before releasing a statement to an SP to adhere to privacy laws.

- **Requirement 3 - Correctness of identity information:** It has to be enforced that the outsourced statements are not manipulable by any party except the HO to prevent the IdPR, the user, and other SPs from manipulating access control decisions of an SP by altering statements of specific users.

- **Requirement 4 - Freshness of identity information:** It has to be possible for SPs to check whether the outsourced statements are up-to-date. In particular, this is necessary to prevent the IdPR from "freezing" statements of a user. For instance, by freezing the statement "user A is administrator", the user A still appears to be an administrator to the SP even if the HO has already revoked the according statement.

- **Requirement 5 - Authenticity of users:** The only party that may be able to impersonate the user should be the HO that maintains the user's identity. Thus, it has to be enforced that the IdPR, SPs, and other users do not get hold of the user's credentials or means like password hashes that can be used to derive those credentials.

In some deployment scenarios the attacker model is weaker and not all of the listed requirements have to be satisfied. We show in Section 4.5.8 how the Occasio approach can be tuned to satisfy less strict security requirements in exchange for other quality characteristics.

Since the HO is assumed to be not available when a user accesses the service, a fundamental shift is required with regard to user authentication. Users have to authenticate against the SP rather than the outsourced IdP because the IdPR is not trusted by the SP to authenticate users. In Section 4.5.5, we discuss the drawbacks that are implied by authenticating users against the SP instead of the HO and how they can be addressed.

4.5.2 Overview of Occasio's Architecture

To address the problem setting of secure identity outsourcing, we propose the Occasio approach. Occasio makes use of security mechanisms to enforce the requirements that we introduced in Section 4.5.1. In the following, we first show how Occasio can address all listed requirements via security mechanisms. Later, we investigate the costs that come with each security mechanism that is required to satisfy strict security requirements and show how Occasio can be tuned for deployment scenarios that contain less strict security requirements.

An overview of Occasio is shown in Figure 4.7. To outsource a user's digital identity, the home organization encrypts the statements on the user's identity attribute values and outsources them to the IdPR along with a so-called proof of freshness (PoF) that can be used to verify that the statements are correct and up-to-date (step 1).

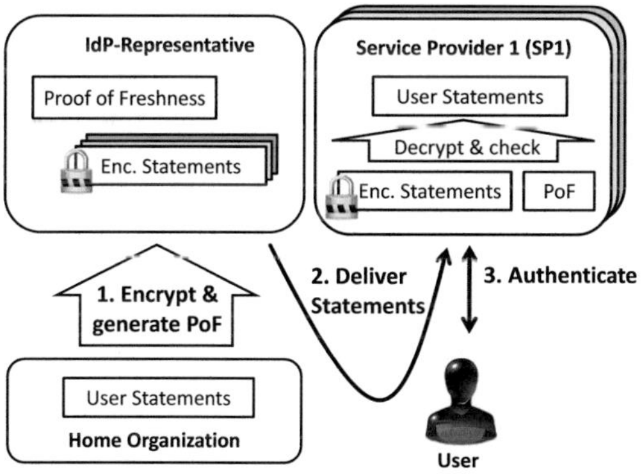

Figure 4.7: Overview of the Occasio concept.

When a user wants to access an SP, the outsourced statements and the PoF are retrieved from the IdPR and forwarded to the SP where they are decrypted and checked for correctness as well as freshness based on the PoF (step 2). In step 3 the user authenticates herself against the SP.

We introduce the security mechanisms that Occasio can apply to enforce the security requirements of a given deployment scenario in the following sections. We show how the confidentiality of statements (and, thus, user identity attributes) can be enforced in Section 4.5.3. In Section 4.5.4 we show how Occasio can enable SPs to check the correctness and freshness of statements. In Section 4.5.5 we introduce the options that Occasio brings to authenticate users. We evaluate how those security mechanisms can be integrated with the existing SAML standard to improve integrability in Section 4.5.6 and investigate the efficiency impacts of the security mechanisms in Section 4.5.7. We show how the negative effects that are induced by security mechanisms can be reduced by only selectively applying security mechanisms to satisfy deployment scenario requirements in Section 4.5.8. As Occasio is able to selectively apply security mechanisms depending on the deployment scenario's security requirements, Occasio constitutes a tunable approach.

4.5.3 Enforcement of the Confidentiality of Identity Information

To ensure that only authorized SPs can read the outsourced statements, the HO can use public-keys that belong to the authorized SPs to encrypt the statements. Public-keys that can be verified to belong to a federation participant exist in many federated identity management standards to enforce trust relationships on a technical level [HCH+05]. For instance, for each participant of a SAML federation, a public-key is contained in the metadata file (see Section 4.4.1).

To outsource the statements $A_{U_1,1}, A_{U_1,2}, \ldots, A_{U_1,k}$ on a user U_1 for an SP S_1, the HO uses the public-key PK_{S_1} of SP S_1 to encrypt the statements[15]:

$$E_{U_1,S_1} = Enc_{PK_{S_1}}(\{A_{U_1,1}, A_{U_1,2}, \ldots, A_{U_1,k}\})$$

The encrypted statements $E_{U_1,S_1}, \ldots, E_{U_n,S_1}$ of the users $U_1, \ldots, U_n$ can be outsourced to the IdPR without jeopardizing the confidentiality of identity information. The IdPR sends the encrypted statements to the SP just like an IdP would send plaintext statements. The SP can then use its private key to decrypt the encrypted statements and make an authorization decision based on the decrypted statements. Based on the assumption that the utilized encryption schemes are secure, only the SP that is authorized to view the statements can decrypt the encrypted statements. Thus, the confidentiality of identity information is preserved.

If an identity is updated, the new statements are encrypted again and replace the old outsourced encrypted statements.

[15] In fact, hybrid encryption is used by Occasio to increase performance, i.e., the statements are encrypted symmetrically with a key that is encrypted with the public-key of the SP.

4.5.4 Enforcement of the Correctness and Freshness of Identity Information

To prevent other parties than the HO from manipulating statements and, thus, altering authorization decisions as well as user rights, a trivial approach would be to let the HO use cryptographic signature schemes [GMR88, ElG85] to sign each outsourced statement. Public-keys that can be verified to belong to a federation participant exist in many federated identity management standards to enforce trust relationships on a technical level [HCH⁺05]. However, this approach only ensures correctness. Freshness is not enforced since SPs have no way of noticing whether an attacker replaced up-to-date statements with outdated statements that were formerly signed by the HO.

To guarantee perfect freshness, the HO has to be queried for the up-to-date statements as the IdPR could suppress changes to statements otherwise. Thus, the objectives of freshness and no involvement of the HO in the process of making authorization decisions are conflicting and there exists a trade-off between freshness guarantees and the involvement of the HO. As we show in the following, Occasio can tune this trade-off to fit the requirements of a given deployment scenario.

A naive approach to enforce a certain degree of freshness is to let the HO regularly attest the freshness of statements by issuing *timestamped signatures*, i.e., by signing the statements together with the current timestamp. Based on a timestamped signatures, SPs can decide whether the statements can be considered "fresh" by rejecting statements with a timestamp older than t seconds. In the following we denote the time window of t seconds in which the SP accepts statements as *freshness window*. The freshness window adjusts the trade-off between the availability of SPs and the integrity guarantees for the SPs. On the one hand, if the HO becomes unavailable and no longer updates the timestamped signatures, users can still access the SPs for t seconds since the SPs still accept statements based on the last issued timestamped signature. On the other hand, an attacker is able to inject outdated statements that are no more than t seconds old.

One drawback of the timestamped signatures approach is that the signatures of all outsourced statements have to be refreshed at least every t seconds. If a user tries to access an SP and the signatures of the transmitted statements is older than t seconds, the statements are rejected and the user is denied access. Refreshing the signature of each outsourced statement induces a high computational load for the HO. To remedy this drawback Occasio makes use of Merkle Hash Trees (MHTs) that allow the HO to attest the freshness of *all* outsourced statements by updating a single signature.

Parts of the following explanation on how Occasio makes use of MHTs is quoted verbatim from our original publication of Occasio in [KH13]. An exemplary binary MHT is shown in Figure 4.8. Each node $T_{i,j}$ of the MHT contains the hash of the concatenated children nodes' content. Notice that Occasio builds the MHT based on *plaintext* statements on each user and each tree leaf consists of the hash of the plaintext statements $A_{U_i,1}, \ldots, A_{U_i,k}$ of a user. Outsourcing the MHT does not undermine the confidentiality of the outsourced identity information, as the MHT only contains hash values of the plaintext statements and a secret random salt is applied to each statement before hashing to prevent dictionary attacks against the hash values of the leaves. If the IdPR learns the salt, it can check whether a specific statement has been made on a user by hashing the statement together with the salt. To prevent the IdPR from learning the salt, it is encrypted and stored at the IdPR just like the statements themselves. Thus, an attacker cannot use the hash values that are contained in the MHT to derive plaintext identity information and the MHT can be securely outsourced without jeopardizing the confidentiality of the identity information. As shown in [Mer89], it is not possible to tamper with any node without changing the value of the root node. Therefore, for the HO, it is sufficient to sign the root node's value together with a timestamp to acknowledge that all contained statements are valid at a given point in time.

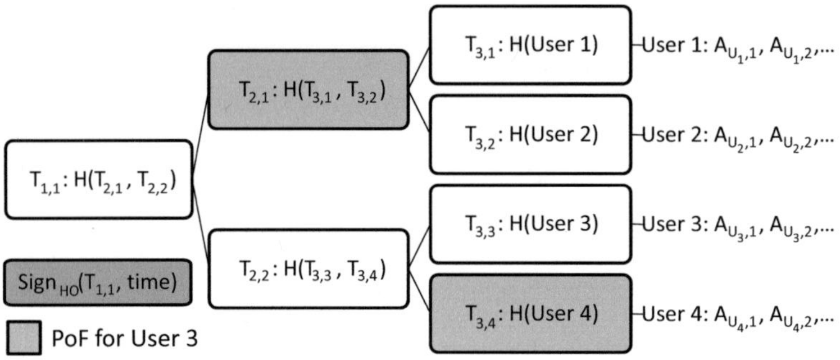

Figure 4.8: Exemplary Merkle Hash Tree [KH13].

Construction of the PoF based on the MHT: The MHT is stored at the IdPR. Thus, using the MHT, the IdPR is able to construct the PoF for a single user *A* as follows:

1. Build the initial PoF by concatenating the siblings of the user *A*'s leaf node.
2. Move on to the parent node and concatenate the siblings of it to the PoF. Repeat until the root node is reached.
3. Add the timestamped signature of the root node to the PoF.

For instance, the PoF of user 3 in Figure 4.8 is $(T_{3,4}, T_{2,1}, Sign_{HO}(T_{1,1}, time))$.

Verification of the PoF by the SP: In order to verify the PoF for the statements $A_{U_i,1}, \ldots, A_{U_i,k}$ of user U_i, an SP performs the following operations:

1. Compute the value of the user U_i's leaf node by hashing the plaintext statements $A_{U_i,1}, \ldots, A_{U_i,k}$ together with their salts.
2. Compute the value of the parent node P_1 by hashing the computed value of the leaf node with its siblings (which are contained in the PoF).
3. Compute the value of P_1's parent node by hashing P_1 with its siblings that are contained in the PoF. Repeat this step until the value of the root node is computed.
4. Verify the signature of the root node's value and check whether the timestamp lies within the freshness window.

To verify the PoF $(T_{3,4}, T_{2,1}, Sign_{HO}(T_{1,1}, time))$ for the statements $A_{U_3,1}, A_{U_3,2}, \ldots$ of user 3, compute $T_{3,3}$ by hashing the statements. Afterwards, compute $T_{2,2}$ by hashing $T_{3,3}$ with $T_{3,4}$ from the PoF. Finally, compute the value $T_{1,1}$ by hashing $T_{2,2}$ with $T_{2,1}$, verify the signature $Sign_{HO}(T_{1,1}, time)$ and check whether timestamp $time$ is within the freshness window.

When the HO outsources statements of a new user or updates/deletes statements of an existing user, the HO has to update the MHT to reflect these changes to the outsourced statements. To prevent attackers from manipulating statements when the HO updates the MHT, the HO has to make sure to perform the updates on the most current version of the MHT and the IdPR must not be able to feed the HO any manipulated or outdated MHT nodes. For the sake of simplicity, we assume that in addition to the outsourced MHT at the IdPR the HO stores a local copy of the MHT to avoid such attacks. However, in principle it would suffice if the HO only stores a copy the current root hash of the MHT which can be used to check whether the rest of the MHT is up-to-date.

Update of existing statements: If the HO updates statements $A_{U_i,1}, \ldots, A_{U_i,k}$ of the user U_i to $A'_{U_i,1}, \ldots, A'_{U_i,k}$, the HO has to update the MHT as follows:

1. Retrieve the sibling nodes of the nodes that are on the path of user U_i's leaf node to the root node.
2. Compute the new value of user U_i's leaf node by hashing the new plaintext statements $A'_{U_i,1}, \ldots, A'_{U_i,k}$ together with their salts.
3. Compute the value of the parent node P_1 by hashing the computed value of the leaf node with its siblings (which are contained in the PoF). Repeat this step until the value of the root node is computed.
4. Compute the signature of the root node's value and the current timestamp.
5. Update the changed nodes of the MHT and the root node signature at the IdPR.

The old root signature is no longer valid for the updated MHT. Thus, the IdPR is not able to suppress the update beyond the freshness window. The following example illustrates the process of updating a user's identity.

> *Example:* Consider the MHT that is shown in Figure 4.8. To outsource new statements $A'_{U_3,1}, \ldots, A'_{U_3,k}$ for user 3, the HO has to retrieve the sibling nodes ($T_{2,1}$ and $T_{3,4}$, marked gray in Figure 4.8) of the nodes that are on the path of user U_3's leaf node to the root node. Then, the HO computes the new value of user 3's leaf node $T_{3,3}$ by hashing the new user statements $A'_{U_3,1}, \ldots, A'_{U_3,k}$ together with their salts. The new values for $T_{2,2}$ can be computed by hashing the computed value of user 3's leaf node $T_{3,3}$ together with the retrieved node $T_{3,4}$. The new value of the root node $T_{1,1}$ can be computed by hashing the computed value of $T_{2,2}$ with the retrieved value of $T_{2,1}$. Finally, the HO signs the new value of the root node $T_{1,1}$ together with the current timestamp and sends the new root signature as well as the updated MHT nodes to the IdPR.

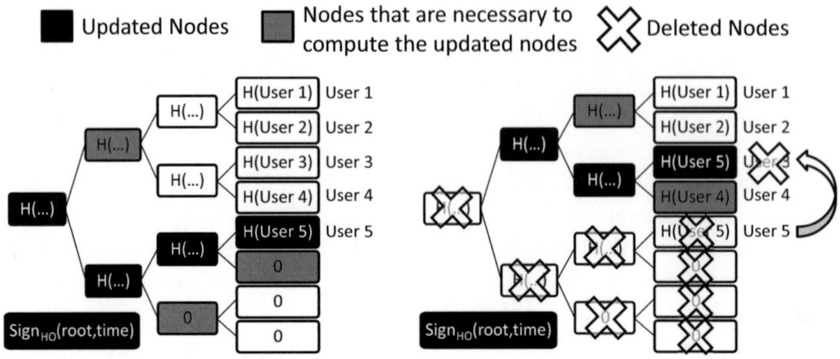

(a) Adding a new user.

(b) Deleting an existing user.

Figure 4.9: Examplary modifications to the Merkle Hash Tree when adding and deleting users.

Adding new users: If the HO outsources statements $A_{U_i,1}, \ldots, A_{U_i,k}$ for a new user U_i, the MHT has to be updated by the HO as follows:

1. First, a leaf node has to be assigned to the user. For efficiency reasons, our goal is that the MHT remains balanced. Thus, we check if a leaf of the MHT is unoccupied, i.e., no statements on another user are stored in that leaf. If a leaf is unoccupied, we store the statements of the new user there. If no leaf node is unoccupied, we create a new root node and declare the old root node to be a child of it. All nodes of the new, empty branch are initialized with the value 0 and the leaf nodes are marked as unoccupied. Thus, given that the old MHT contained n leaf nodes, n leaf nodes are unoccupied in the new MHT.

2. Compute the hash value of user U_i's statements by hashing the new plaintext statements $A_{U_i,1}, \ldots, A_{U_i,k}$ together with their salts. We store the hash value of U_i's statements in the first unoccupied leaf.

3. Retrieve all sibling nodes of the nodes that are on the path from the leaf node that is assigned to the new user U_i to the root node of the MHTs.

4. Compute the value of the parent node P_1 by hashing the computed value of the leaf node that contains the statements of U_i with its siblings which have been retrieved. Repeat this step until the value of the root node is computed.

5. Compute the signature of the computed root node's value and the current timestamp.

6. Update the changed nodes of the MHT and the root node signature at the IdPR.

The old root signature is no longer valid for the updated MHT. Thus, the IdPR is not able to suppress the insertion of new users beyond the freshness window. The following example illustrates the process of adding a new user identity.

> *Example:* Consider the MHT that is shown in Figure 4.8. Figure 4.9a illustrates the process of updating the MHT when adding a new user 5. As the original MHT has no room for additional users, a new root node is created which, in the new MHT, constitutes the parent of the original root node. All new nodes are initialized with 0. To insert the new user 5, all siblings of the nodes on the path from the first unoccupied leaf node to the root node are retrieved (dotted nodes in Figure 4.9a). Based on these nodes the new values of the nodes on the path from user 5's leaf node to the root node (striped nodes in Figure 4.9a) can be computed. Finally, the new root node is signed and the updated nodes as well as the new root signature are outsourced to the IdPR.

Deletion of existing users: If the HO deletes a user U_i, the HO has to update the MHT as follows:

1. For efficiency reasons we aim to balance the MHT and reduce the height of the MHT as much as possible. To achieve this, we guarantee that the MHT contains no "gaps" in the sense that there is an unoccupied leaf node between two occupied leaf nodes. To avoid leaving a gap by removing user U_i, swap the last occupied leaf node with the leaf node of user U_i, set the contents of the last occupied leaf node to 0, and mark it as unoccupied.

2. Check whether only half of the MHT leaf nodes are occupied. If this is the case, reduce the height of the MHT by one level declaring the child of the root node that contains leaf nodes with user statement hashes as the new root node.

3. Let us assume that the last occupied leaf node in the MHT belongs to user U_n. As the values of two leaf nodes changed, both the values of the nodes on the path from the former U_i's leaf node to the root node and from the former U_n's leaf node to the root node have to be updated. Retrieve all sibling nodes of the nodes that are on the path from user U_i's leaf node to the root node of the MHTs. Also retrieve all sibling nodes of the nodes that are on the path from user U_n's leaf node to the root node of the MHTs. Based on the new value of each leaf node and these retrieved values, the values of the nodes on the path from the leaf nodes to the root node can be computed.

4. Compute the signature of the root node's value with the current timestamp.

5. Update the changed nodes of the MHT and the root node signature at the IdPR.

The old root signature is no longer valid for the updated MHT. Thus, the IdPR is not able to suppress the deletion of a user beyond the freshness window. The following example shows the process of deleting a user's identity.

> *Example:* Consider the MHT that is shown in Figure 4.9a. Figure 4.9b illustrates the process of updating the MHT when removing user 3. First, the last leaf node – which belongs to user 5 in this case – is swapped with

the leaf node of the user 3 who is deleted. As the lower branch from the root node of the MHT in Figure 4.9b now only contains unoccupied leaf nodes, it can be discarded. The root node of the new MHT is the child of the old root node that still contains branches with occupied leaf nodes. As the third leaf node now contains user 5 instead of user 3, the path from user 5's leaf node to the root node has to be updated. Thus, all siblings of the nodes on the path are retrieved (dotted nodes in Figure 4.9b). Based on the sibling nodes, new values can be computed for the nodes on the path from user 5's node to the new root node (striped nodes in Figure 4.9b). The resulting new root node is then signed by the HO and the signature is outsourced to the IdPR together with the updated MHT nodes.

4.5.5 Enforcement of User Authenticity

Since the HO is assumed to be not available when users access the service, a fundamental shift is required with regard to user authentication. Users have to authenticate against the SP rather than the outsourced IdP because the IdPR is not trusted by the SP to authenticate users. User authentication against the SP has two drawbacks which is why it is avoided in traditional federated identity management: 1) the password can be intercepted at multiple SPs which increases the attack surface and 2) each SP has to manage verification tokens such as password hashes to validate the passwords of a user. In this section we show how both drawbacks can be addressed.

Occasio provides multiple ways to authenticate users. Each authentication method has advantages and disadvantages over the others and it depends on the credential trust model of the federation (see Section 4.2.2) whether the authentication method can be applied in a given deployment scenario.

Full-trust credential model: In the full-trust credential model, SPs of the federation are allowed to view user credentials. To leverage that, Occasio allows the home organization to encode user password hashes in statements. These user password hashes are encrypted and outsourced to the IdPR like regular statements. When a user tries to access a service of the SP, the encrypted statements are retrieved from the IdPR, sent to SP, and decrypted there. To authenticate, the user passes her password to the SP which checks whether it matches the decrypted password hash.

Limited-trust credential model with home-organizational credentials: In the limited-trust credential model, SPs are not allowed to view user credentials as they are assumed to be possibly compromised and must not be able to use user credentials to impersonate users and access other SPs. Thus, giving them access to password hashes is not an option. As the SP does not trust both the potentially malicious IdPR to perform an authentication, the SP itself has to authenticate its users. Occasio makes use of challenge-response protocols based on public-private key pairs that allow the SP to authenticate users based on their public-keys without having the opportunity to derive the private key from the process of authentication. The home organization outsources statements that include the users' public-keys. When logging in, the public-key statements are retrieved from the IdPR, sent to the SP, and decrypted. Based on the contained public-key, the SP can authenticate the user. Compared to using Occasio in the full-trust credential model, usability can be affected since the users cannot use passwords for authentication but have to manage cryptographically strong private keys on their end-devices from which they want to access the SPs.

Limited-trust credential model with service local credentials: Requiring the users to manage cryptographically strong keys can be prevented in the limited-trust credential model by allowing the users to establish SP-specific passwords with the SPs. By doing this, the federated identity management advantage of requiring the user to only remember a single credential to log on to various SPs is forfeited. However, in practice many users prefer remembering multiple passwords instead of synchronizing cryptographic keys on their devices [HVO12].

We discuss the usability challenges in Section 4.5.9 and show how they can be mitigated.

4.5.6 Integration with SAML

In this section we show how Occasio can be integrated with the SAML standard [HCH⁺05] to minimize integration effort and achieve interoperability with existing SAML-based SPs. SAML is widely used to federate web-based services and can also be used to federate non web-based services as we showed in Section 2.4. By integrating Occasio with the SAML standard, many use cases that are already covered by SAML can be addressed. Furthermore, existing extendable SAML frameworks such as Shibboleth[16] and simpleSAMLphp[17] which are widely deployed already can be leveraged to implement Occasio. In the following we show how we integrated Occasio with SAML conceptually and implemented Occasio as a simpleSAMLphp extension.

Conceptual integration: In the following, we show how Occasio's security mechanisms can be integrated with the SAML standard on a conceptual level, independent from specific IdP and SP implementations.

Confidentiality and correctness: SAML supports the encryption of statements using so called *SAML Encrypted Attribute Statements*. The HO builds SAML-Assertions that include pre-encrypted SAML Encrypted Attribute Statements, signs, and outsources them to the IdPR. The IdPR can send these pre-encrypted and pre-signed SAML-Assertions to the SPs upon request. As the outsourced statements are regular SAML-Assertions, regular SAML SPs can process them.

Freshness: If attribute freshness is required in the given deployment scenario, the SAML SPs have to be able to receive and validate PoFs. Validation of MHT-based PoFs is beyond the scope of the SAML standard and has to be implemented additionally at the SP. However, SAML can be used to convey the PoF from the IdPR to the SAML SP by encapsulating the PoF into a dedicated SAML Attribute Statement. Thus, SAML-Protocols, SAML-Bindings, and SAML-Profiles do not need to be modified and can be

[16] http://shibboleth.net/ [last visited on March 2015]

[17] https://simplesamlphp.org/ [last visited on March 2015]

used in the accustomed way. For instance, our simpleSAMLphp extension leverages the SAML-WebSSO profile that allows web-based services to request statements on a user by redirecting her to the IdPR. Upon being provided the user's username, the IdPR issues the according stored SAML-Assertions that also contain the PoF and automatically redirects the user back to the SP.

Authenticity: In absence of a party that is trusted by the SP to authenticate users the SP itself has to authenticate the users[18]. SP-based authentication is beyond the scope of the SAML standard and SAML SPs potentially have to be modified to support it. However, SAML can be used to convey encrypted credentials from the IdPR to the SP as explained in Section 4.5.5. SAML-Protocols, SAML-Bindings, and SAML-Profiles can be used in the accustomed way and no adjustments to the SAML protocol are required.

Implementation: In the following, we show which adaptations to the SPs and IdPs are required in practice based on the example of our simpleSAMLphp Occasio extension. The architecture of the implementation of our simpleSAMLphp Occasio extension is shown in Figure 4.10. By extending the simpleSAMLphp framework which already contained most of the required functionality we increase code maintainability. Our implementation consists of the so-called IdP-component that is run by the IdPR and the SP-component that is run by the SP.

Confidentiality and correctness: We adapted the simpleSAMLphp IdP-component to retrieve the pre-signed and pre-encrypted SAML-Assertions from a database. As the pre-signed and pre-encrypted SAML-Assertions are regular SAML-Assertions, the SP-component does not have to be modified in any way.

Freshness: If attribute freshness is required by an SP, a SAML SP that supports MHT-based signatures is required. To our knowledge this is not supported by any SAML SP implementation to date. We extended the simpleSAMLphp framework with this functionality. We added the support to encapsulate the PoF into an Attribute Statement in the IdP-component and to verify the encapsulated PoF in the SP-component. Note that we did not alter the interface of the simpleSAMLphp SP-component. Thus, existing services that rely on the simpleSAMLphp SP implementation do not have to be modified even if our extended simpleSAMLphp SP-component is required to provide freshness guarantees.

Authenticity: In case the SP may view the users' password hashes, our simpleSAMLphp extension checks if the password sent by the user matches the password hash that is contained in the outsourced statements. In case an SP-specific password was previously set by the user, our extension checks if the password that the user provides for authentication matches the previously stored password. To authenticate the user via challenge-response (see Section 4.5.5) our extension uses the user's public-key that is contained in an outsourced statement to perform a client certificate based authentication during the TLS handshake in the SAML-WebSSO procedure.

Trade-off between compatibility and freshness/user authentication: We integrate Occasio with SAML in such a way that unmodified SAML SPs are compatible

[18] The HO is trusted to authenticate users but does not participate during user log ins (see Section 4.5.1).

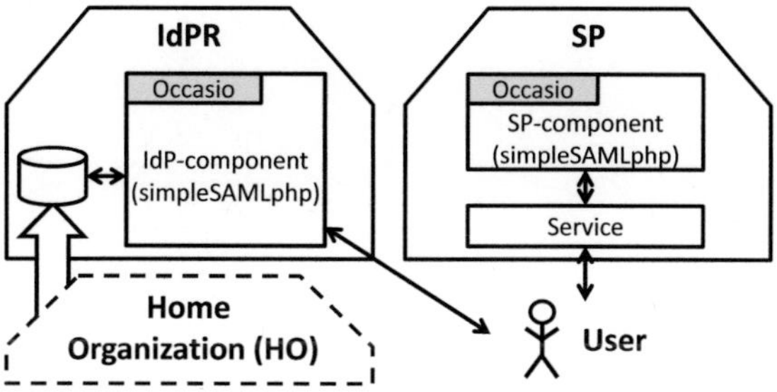

Figure 4.10: Occasio implementation based on simpleSAMLphp [KH13].

to Occasio IdPs if 1) no freshness guarantees are necessary and 2) the service does not rely on the SAML SP-component to authenticate users. If freshness is required, SPs have to support MHT-based signatures. If services rely on the SP-component to authenticate users, standard SAML SPs cannot be used as they rely on the IdP to authenticate users. Occasio can be tuned to support standard SPs by forfeiting freshness guarantees and user authentication. We discuss how the impact of these compatibility drawbacks can be further reduced in Section 4.5.9.

4.5.7 Performance Evaluation

We measured the performance of our simpleSAMLphp extension to analyze the efficiency costs that are induced by applying Occasio's security mechanisms. We measured the performance of logins based on Occasio and the efficiency of applying changes to the outsourced identity data. The overhead for frequently updating the MHT's root signature is negligible, as only a single signing operation has to be performed. Our measurements were conducted in a testbed that consists of an IdPR machine (QuadCore 2.50GHz, 4GB RAM), an SP machine (OctoCore 2.00GHz, 4GB RAM), a HO machine (DualCore 2.50GHz, 4GB RAM) and a user machine (DualCore 2.93GHz, 4GB RAM).

To measure the **performance of logins**, i.e., the process of user authentication and authorization, we deployed our simpleSAMLphp extension on the IdPR and the SP machine. The IdPR stores encrypted statements for the outsourced identities, the MHT, and the signature of the MHT's root in a PostgreSQL database[19]. We measured the latency and the throughput of SAML WebSSO based logins. The logins that we measured are performed as follows: after requesting access to a service of the SP, the user is redirected to a discovery service where she chooses her identity provider (in case of Occasio, the IdPR). At the IdPR, she enters her username, the IdPR returns her encrypted statements which also include the PoF and she passes them to the SP. The SP decrypts the statements, checks the PoF for validity, and authenticates the user via client certificate (see Section 4.5.5). As the measured login process includes every security mechanism available to Occasio, the conducted

[19] http://www.postgresql.org [last visited on March 2015]

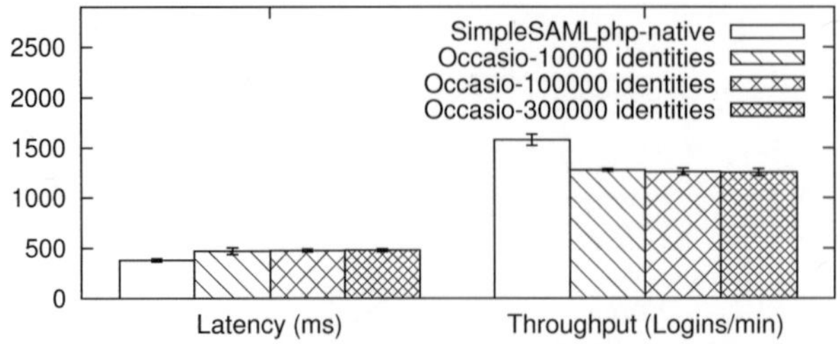

Figure 4.11: Performance of the login process (95% confidence interval) [KH13].

measurements are worst-case measurements[20], i.e., Occasio performs better if not all security mechanisms are utilized.

The average latency and throughput of 100000 concurrently executed login procedures for randomly chosen users is shown in Figure 4.11. To compare our results against a baseline, we also measured the performance of an unmodified simpleSAMLphp deployment (IdP and SP) in our testbed. For the native simpleSAMLphp measurements we repeatedly logged in one single hard coded user. No overhead for retrieving identities from backend identity management systems like LDAP servers is induced for the IdP. Thus, the measurements constitute an upper bound for the performance of the native simpleSAMLphp performance.

Occasio's login performance overhead is negligible: *The results show that Occasio does induce a slight overhead due to the decryption of statements and the validation of the PoF. However, the overhead is negligible as it can be compensated with slightly more performing hardware. Furthermore, the results show that the number of outsourced identities does not have a visible influence on the performance of logins. Thus, the measurements indicate that Occasio scales well in terms of the number of maintained identities.*

To measure the efficiency of **applying changes to the outsourced identities** we let the HO issue requests to change the outsourced statements, update the MHT, and the root signature. We made use of SQL transactions to avoid inconsistent PoFs for users that log on while changes are made to the MHT. To gain more insight on the extent of each security mechanism's overhead, we additionally measured the isolated performance of the HO, i.e., without updating anything in the database. Furthermore, we measured the isolated performance of the IdPR, i.e., the performance of just storing encrypted statements and the MHT in the database. We measured the performance for 10000, 100000, and 300000 identities for 1, 5, and 10 SPs. The measured throughput rates in changes per minute (C/m) for 10000 sequentially conducted changes on randomly selected identities are shown in Table 4.1.

The measurements show that both the number of managed identities and the number of SPs influence the performance of applying changes. The performance decline for more SPs can be explained by the fact that the statements have to be encrypted separately for each SP. The measurements of the HO's performance confirm this

[20] Authentication via password is not covered. Yet, we argue that verifying a password is more efficient than an authentication via client certificate. Thus, our measurments can be seen as worst-case measurements.

Managed identities	Real scenario: Throughput (95% confidence interval)		
	1 SP (C/m)	5 SPs (C/m)	10 SPs (C/m)
10000	3178 (±6)	2825 (±3)	2012 (±1)
100000	2924 (±15)	2821 (±1)	2010 (±1)
300000	2758 (±16)	2737 (±9)	2005 (±1)

Overall performance

Managed identities	HO component's performance, no database updates		
	1 SP (C/m)	5 SPs (C/m)	10 SPs (C/m)
10000	4529 (±2)	2952 (±1)	2072 (±1)
100000	4524 (±2)	2951 (±1)	2067 (±1)
300000	4517 (±2)	2947 (±1)	2067 (±1)

HO's performance

Managed identities	Only database updates at the IdPR component		
	1 SP (C/m)	5 SPs (C/m)	10 SPs (C/m)
10000	5464 (±19)	5452 (±7)	5436 (±14)
100000	4862 (±10)	4854 (±10)	4851 (±9)
300000	4418 (±18)	4412 (±29)	4400 (±22)

IdPR's performance

Table 4.1: Changes per minute (C/m) for the real scenario, for encrypting the values at the HO, and for storing them at the IdPR [KH13].

explanation, as these measurements are only influenced by encryption and signing operations of the HO. The throughput of the IdPR is not affected by more SPs as for each additional SP just a single additional UPDATE operation has to be performed to update the encrypted statements. As the MHT is build on plaintext attributes rather than encrypted statements, just a single MHT needs to be updated. Thus, the overhead for updating the MHT does not depend on the number of SPs.

Trade-off: higher efficiency from the HO's perspective if confidentiality is not required: *The measurements show that applying encryption on statements induces efficiency overhead for the HO. We argue that the measured overheads are not critical as the number of identity changes per minute are not in the order of 1000s in many deployment scenarios. Nevertheless, the overhead can be saved by only enforcing confidentiality if necessary. If encrypted statements are not required, plaintext statements could be outsourced. These plaintext statements would not have to be outsourced for each SP separately, but just once.*

The throughput measurements of the HO component are nearly independent from the number of managed identities. The throughput measurements of the IdPR, however, seem to depend on it. Only the MHT, the root signature and the statements in the database are updated in the isolated IdPR measurements. For each updated identity, the root signature needs to be updated and the statements of the identity have to be updated once. To update the MHT, $\lceil log(n) \rceil$ nodes have to be updated where n is the number of outsourced identities. This logarithmic dependency on the number of outsourced identities is reflected in the throughput measurements.

"When removing users, a 'gap' is induced into the MHT. To fill this gap, the last user is placed at the position of the removed user. Thus, additional $\lceil log(n) \rceil$ nodes of the MHT have to be updated. To sum up, it is to be expected that the performance of updating existing users matches the performance of adding new users. Furthermore, the throughput of the IdPR component for removing users is expected to be half of the throughput for updating existing users, as the double amount of MHT nodes has to be updated. However, the throughput of the HO component can be expected to be significantly higher, as no encryption needs to be performed to remove a user." [KH13]

Trade-off: higher efficiency from the IdPR's perspective if freshness is not required: *The measurements show that enforcing freshness via MHTs has a negative impact on the efficiency of the IdP-component. However, we argue that the achieved throughput of changes per minute is sufficient in most deployment scenarios. Regardless, the overhead for the IdPR can be saved by only making use of MHTs if freshness has to be enforced.*

"Notice that we measured Occasio on a specific testbed. In terms of the computational overhead for encryption and signing, the performance of both login processes and applying changes can be increased by using more performing hardware or multiple load-balanced instances of the IdPR, the HO, or the SP components. Furthermore, the potential IO-bottleneck of the underlying database can be omitted by partitioning the identities to multiple databases. In turn, the MHT has to be partitioned as well and the HO has to regularly refresh the root of each MHT. The signing of a reasonable number of MHT roots constitutes a negligible overhead." [KH13]

Overall, the measurements showed that outsourcing digital identities based on MHTs is efficient enough to be applied in practice. Monitoring the rate of changes that are applied on identities of the Karlsruhe Institute of Technology showed that no more than 100 changes per minute were applied to the 30000 managed identities even in peak times [Hö11].

4.5.8 Tunability of Occasio's Security Characteristics

As shown in Section 4.5.4, Occasio allows to tune the inherent trade-off between freshness and availability of identity information by adjusting the required freshness window. Thus, depending on the freshness requirements of the deployment scenario, Occasio optimizes availability of services that are provided by SPs.

Furthermore, Occasio can be tuned to omit specific security mechanisms in favor for other quality characteristics. Depending on the deployment scenario, enforcing all security characteristics that can be enforced by Occasio is unnecessary. This can be due to the federation trust model (cf. Section 4.2.2) or due to the fact that it is not worthwhile to enforce specific security characteristics from a risk management perspective. For instance, it makes little sense to protect the confidentiality of identity information that is publically accessible via other channels. In the following we show how other quality characteristics can be improved by not enforcing specific security characteristics and exemplarily list scenarios in which it is feasible to selectively omit the application of Occasio's security mechanisms. An overview of the trade-offs between security characteristics and other quality characteristics is shown in Figure 4.12.

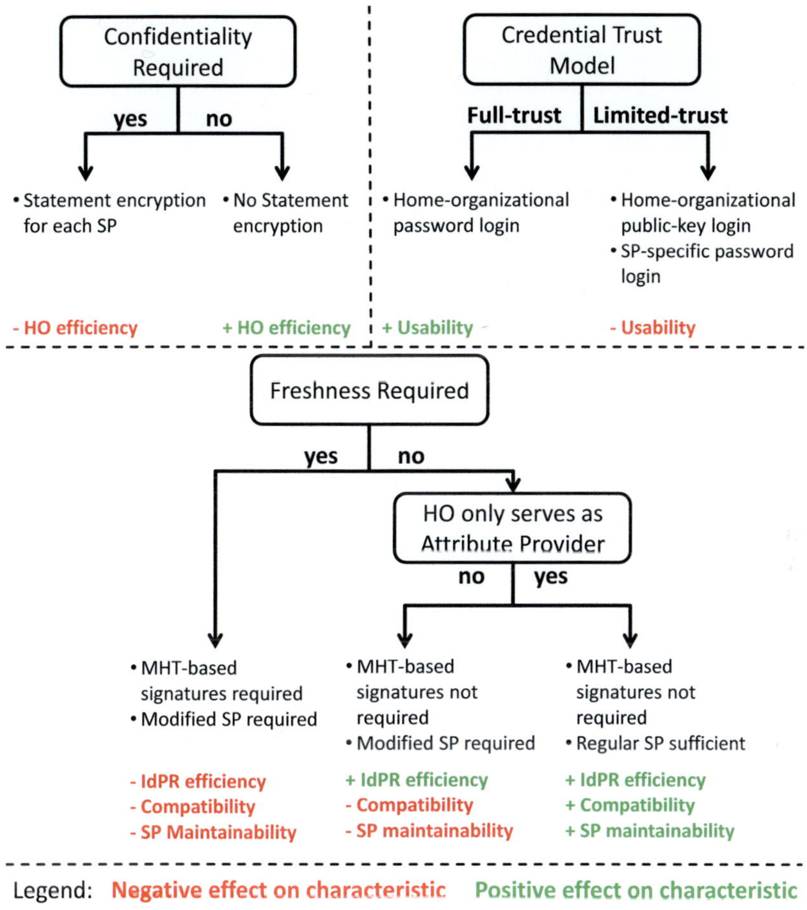

Figure 4.12: Tunable characteristics of Occasio.

Confidentiality Requirements: If confidentiality of the outsourced identity information has to be enforced, each identity statement has to be encrypted for each SP. Encrypting the statements just once with an encryption key that is shared with all relying SPs is not possible in many cases as not every SP should be able to view every identity statement[21]. As shown in Section 4.5.7, the encryption of statements for each SP incurs a substantial efficiency overhead for the HO when applying changes to the outsourced identities. If confidentiality of the outsourced identity information does not have to be enforced, the efficiency overhead for the HO can be avoided and more changes to the outsourced identities can be made in the same time.

As an example in which confidentiality does not need to be enforced consider the following deployment scenario.

[21] In many deployment scenarios, the user has to give her consent for the release of identity information to a specific SP [GSKK07].

Example: Enforcing the confidentiality of identity information is not required if the information is publically accessible anyways. For instance, friend relationships within social networks are often publically visible. Today many applications exist that make authorization decisions based on Facebook friend relationships. For instance, an application can enforce that only a user's friends may be able to view the posts of the user. Thus, it is important to enforce the freshness and correctness but not the confidentiality of the friendship relationships.

Credential Trust Model: As shown in Section 4.5.5, Occasio allows users to authenticate via their home-organizational password as it is usual in the federated identity management scenario if the SPs are trusted to view the user credentials (full-trust credential model). Thus, the user only has to remember one password to access multiple services that are provided by multiple organizations.

If the SPs are not trusted to view credentials (limited-trust credential model), Occasio offers the users to authenticate to the SP via challenge-response protocols as well as previously established SP-specific passwords. Relying on previously established SP-specific passwords requires the user to memorize one password for each SP which also constitutes a usability hurdle and undermines one of the major benefits of federated identity management. Challenge response protocols can be used to avoid this. They are widely used in some deployment scenarios such as SSH access via a deployed public-key. The user answers a challenge of the SSH server based on his cryptographically strong private key to authenticate. Requiring the user to authenticate via challenge-response protocols does not have a significant effect on usability if the private keys are already present on the users end devices and the user is accustomed to rely on challenge-response protocols. For instance, if Occasio is used to federate access to an SSH service, users that already use SSH public-key based authentication will hardly notice a difference. Using challenge-response protocols to access web-based services is possible but not commonly used. Thus, relying on challenge-response protocols for authentication in this setting can constitutes a usability hurdle.

Example: A real-world example for a federation with the full-trust credential model is the bwIDM federation [KLS+14]. A real-world example for a federation with the limited-trust credential model is the DFN-AAI federation (see Section 4.2.2).

Freshness Requirements: If freshness is required in a deployment scenario, Occasio makes use of the MHT-based signature that we introduced in Section 4.5.4 to reduce the efficiency overhead induced for the HO when attesting the freshness of all outsourced identities. To enforce freshness, an SP has to install a modified SP implementation that can validate MHT-based signatures as MHT-based signatures are not supported by most SP implementations[22]. Furthermore, the maintenance of the MHT constitutes efficiency overhead for the IdPR as we showed in Section 4.5.7.

22 We refer to SAML SP implementations. However, the statement is also true for SP implementations that implement other standards such as OAuth.

If freshness is not required in a given deployment scenario, MHT-based signatures are not required and can be omitted. As no MHT has to updated when identities are updated, updates can be processed by the IdPR more efficiently. However, if the SP relies on the HO not only for providing identity information but also to authenticate users, a modified SP implementation is still required in many cases as the user has to be authenticated at the SP. In traditional federated identity management, the user authenticates to the HO and the HO asserts that the user successfully authenticated to the SP. Thus, most existing SP implementations do not support user authentication and a modified SP implementation like our simpleSAMLphp extension is required.

If freshness is not required in the deployment scenario and an SP relies on the HO only for identity information, regular, unmodified SP implementations are compatible with Occasio IdPs (see Section 4.5.6). Operating standard non-modified SP implementations increases maintainability. For instance, patches to the software can be easily applied without having to check for conflicts with non-standard modifications. Furthermore, compatibility is increased as integrating the IdPR with existing SPs that are already up and running is possible without requiring the SPs to modify their SP implementations.

Consider the following deployment scenarios as an example in which freshness does not need to be enforced:

> *Example:* In some deployment scenarios, the SPs do not rely on authorization tokens that are managed by the home organization of the user (see Section 4.2.2) but only attributes which contain identity information that does not have to be strictly up-to-date. For instance, applications like Photobucket[23] allow to import photos that users previously uploaded to Facebook. These photos constitute identity information on the user. Whether the photos are up-to-date is of minor importance. Another example is Umbrella[24] which constitutes an identity system for the users of the European large photon and neutron facilities. Umbrella constitutes a SAML IdP that allows arbitrary users to register for accounts and enter identity data. As the identity data is not used for authorization (the user can edit the data herself) and only contains contact information such as the postal address, enforcing the freshness of it is of minor importance for most SPs.

4.5.9 Discussion

The security mechanisms that Occasio applies to enforce security characteristics can imply that the user has to use unconventional authentication mechanisms and that SPs have to deploy a modified SP implementation. We showed in Section 4.5.8 that depending on deployment scenario, Occasio can be tuned to avoid these drawbacks. Even if in a specific deployment scenario the approach cannot be tuned to mitigate the usability and compatibility drawbacks of Occasio it is possible to address them by

[23] http://photobucket.com/ [last visited on March 2015]
[24] https://umbrellaid.org/ [last visited on March 2015]

applying Occasio in parallel to traditional federated identity management approaches. Occasio then acts merely as a fallback if a user's HO is not available. Thus, a highly available identity and access management can be guaranteed without requiring the HO to operate highly available infrastructure. At the same time, usability is only influenced if the HO is not available and as long as the HO is available, standard SP implementations are supported. Thus, each SP can decide for itself whether the overhead for using a modified SAML SP implementation for increased availability is justified.

Besides relying on the MHT signatures, SPs can additionally enhance the freshness of identity information by memorizing the timestamp of the most recently observed statement that was issued by a HO, i.e., memorizing the last time the HO was definitely available. Subsequent statements are only accepted if the timestamp of PoF is newer than the most recent observed timestamp. Notice that this enhances freshness in most use cases, but provides no strict guarantees. If a user logs in at time t and no other user logged in between t - w (where w constitutes the freshness window) the technique does not increase freshness. Furthermore, even if users continuously log in, the IdPR can prevent the SPs from learning new timestamps by freezing the states of all users. However, in this case the IdPR is no longer able to selectively freeze the state of users.

We motivated Occasio by the need to outsource identities to external parties in order to achieve a high availability. However, outsourcing identities to external parties can also solve other acknowledged problems such as attribute aggregation [CI09]. Attribute aggregation allows to fuse statements issued by multiple HOs that do not necessarily trust each other with regard to confidentiality and integrity of the identity information and authorization tokens. With Occasio, it is possible for multiple HOs to outsource identities to the same IdPR while guaranteeing that neither the IdPR, nor the other HOs can read or tamper with the outsourced identities. Thus, users can retrieve their digital identity that is managed by multiple authoritative sources from a single provider and avoid the overhead of collecting all pieces of the digital identity from multiple providers.

4.6 Conclusions

We introduced Occasio, a tunable approach to securely outsource identity data to highly available external parties. By outsourcing identity data, it can be made highly available without operating a highly available infrastructure. We analyzed possible attacks in identity outsourcing problem setting and identified security characteristics that have to be enforced to prevent those attacks. Furthermore, we identified and assessed security mechanisms that can be used to enforce those security characteristics and showed how they can be integrated with the federated identity management standard SAML.

In particular we identified the following security mechanisms to enforce security characteristics and their impact on other quality characteristics:

- Confidentiality of the identity information: *Encryption schemes* are applied by Occasio to encrypt each identity for each SP that should have access to it. We showed that applying encryption affects the efficiency of applying identity information updates.

- Correctness of the identity information: *Signature schemes* are applied by Occasio to ensure the correctness of the identity information and guarantee that it was not modified. While issuing and validating signatures constitutes a computational effort, our measurements showed that it does not constitute a dominant factor with regard to login and update efficiency.
- Freshness of the identity information: *MHT-based signatures* are applied by Occasio to ensure the freshness of identity information and guarantee that updates on it cannot be suppressed. We showed that MHT-based signatures induce efficiency overheads for the IdPR when updates are performed. Furthermore, existing SP implementations do not support the validation of MHT-based signatures. To integrate the Occasio concept with them, they have to be modified if freshness is required. Thus, if MHT-based signatures are applied, the compatibility of Occasio with existing SPs is negatively affected.
- Authenticity of the user: Due to an absent home organization, Occasio authenticates users against the SP and offers the following mechanisms: *password authentication based on home-organizational password* requires an SP that can be trusted to view user credentials, *challenge-response protocols* and *password authentication based on SP-specific password* do not require trusted SPs but incur a usability overhead, as users have to manage cryptographically strong keys for challenge-response authentication or multiple SP-specific passwords.

The federation trust model has an impact on the deployment scenario requirements. In particular, it determines which security characteristics have to be enforced and, thus, which security mechanisms can be considered sufficient in a deployment scenario. An approach is required to be tunable if it has to be deployed in a variety of federations with different trust models or the requirements of participants within a single federation differ. We showed how Occasio can be tuned to individual deployment scenario requirements by selectively applying only the required security mechanisms. This is possible as the mechanisms can be applied independently from each other. Occasio showed that for tunable approaches it is important to address the security characteristics independently whenever possible so that quality characteristics can be optimized for deployment scenarios with different security requirements. Furthermore, we proposed signatures based on Merkle Hash Trees to enforce freshness. By adjusting the time window in that such signatures are valid, it is possible to tune the inherent trade-off between the guaranteed freshness of attribute values and the availability of the SP's provided services for the users.

Our conducted efficiency measurements strongly indicate that securely outsourcing identities is feasible in practice. For an increased compatibility of identity outsourcing with common federated identity management environments, standards should be extended to support MHT-based signatures in future work. Furthermore, relying on passwords for authentication inherently affects usability if the SP is not trusted to view password hashes. In future work, further research is required on how to increase the usability of other authentication methods such as challenge-response authentication based on cryptographic keys. We provide a first building block for this direction of research in Chapter 5.

5 Credential Outsourcing

As of today, the high usability of services for end-users such as data synchronization services[1] or e-mail providers[2] outweighs concerns of the users with regard to security characteristics such as data confidentiality or data integrity. In principle, the users could address these concerns by relying on cryptographic techniques such as encrypting data or mails before uploading them to the providers. Secret keys that are used for encryption have to be present on any end-device from which the services are accessed. To properly use modern encryption schemes, the secret keys have to be strong, i.e., they have to be hard to guess. For instance, the National Institute of Standards and Technology (NIST) recommends to use keys with entropy of at least 80 bit [Nat12]. The prevalent type of secret that modern users are accustomed to and that can be memorized by humans are passwords. According to a NIST specification, an 8 character password that was generated by a human can be estimated to have entropy of only 18 bit[3]. Thus, user passwords do not provide enough entropy to act as the strong secret that is required to apply cryptographic techniques. To enhance security characteristics via cryptographic techniques, users have to address the problem of managing strong secrets with high entropy across their devices from which they access the services. This usability hurdle constitutes one of the main reasons why end-users often do not apply cryptographic techniques in practice. In this chapter, we propose a credential repository approach that, based on the assumption that an attacker is not able to compromise multiple parties at once, allows to securely trade a

---

1 e.g., Dropbox: http://www.dropbox.com [last visited on March 2015]

2 e.g., Googlemail: http://www.googlemail.com [last visited on March 2015]

3 A study of leaked user passwords shows that the NIST specification does not accurately predict password strength. Even so, a trained password cracking tool was able to crack more than 20% of the leaked 8 character passwords with only 50000 guesses [WACS10].

weak password for a cryptographically strong secret. This facilitates the management of strong secrets from the perspective of the user, as she only has to remember a weak password. In the following we will use the terms *secret* and *credential* interchangeably. Thus, the credential repositories that we present in this chapter can also be considered as secret repositories that store secrets which do not necessarily constitute credentials.

The chapter is structured as follows. We investigate existing approaches that aim to help the user in managing strong cryptographic secrets and show their drawbacks in Section 5.1. In Section 5.2 we introduce the idea of distributed credential repositories to trade weak passwords for strong secrets and state one of the main research questions that have to be addressed when building distributed credential repositories. We provide an overview on related work on that question in Section 5.3. We present our contribution Credis that addresses the research question in Section 5.4. In Section 5.5, we summarize our contributions and conclude the chapter.

Parts of the contributions presented in this chapter have been previously published in [KMH13]. Whenever appropriate, the following sections are quoted from the original publication.

5.1 The Problem of Managing Strong Secrets

There are several existing approaches to aid the user in managing strong secrets. In the following we will provide an overview of them and highlight their current drawbacks.

A trivial approach to the problem of managing strong secrets is to store the secrets on a secure **hardware token** [SBG02] that the user can connect to end-devices on which the secrets are required (see Figure 5.1a). Such a hardware token can for instance be a smartcard, a USB token, or a smartphone. While this approach is used in some enterprise networks with high security requirements, it is hard to roll out this approach for regular users due to several usability drawbacks. The end-devices on which the secrets are needed have to be compatible with the hardware token, i.e., the hardware token needs to provide an interface that is supported by the end-device. For instance, plugging a USB token into a smartphone is not always possible. Furthermore, users have to carry the hardware token on them when they want to access services that require the secrets. Other disadvantages of hardware tokens include that once the token is lost physically, the secrets have to be considered compromised and that – unlike a password – a hardware token needs to be acquired by the user before it can be used.

Another approach is to **store and synchronize the secrets on end-devices** (see Figure 5.1b). This does not require the user to carry a hardware token. However, the user has to keep the secrets synchronized across all of her devices. While services exist that simplify this synchronization[4], the synchronization needs to be triggered a-priori. Thus, the user has to know which devices will be used in the future. Another drawback of such an approach is that once an end-device of the user becomes compromised by an attacker, the attacker has full access to all the secrets that are stored on this end-device.

[4] e.g., Firefox Sync https://wiki.mozilla.org/Services/Sync [last visited on March 2015]

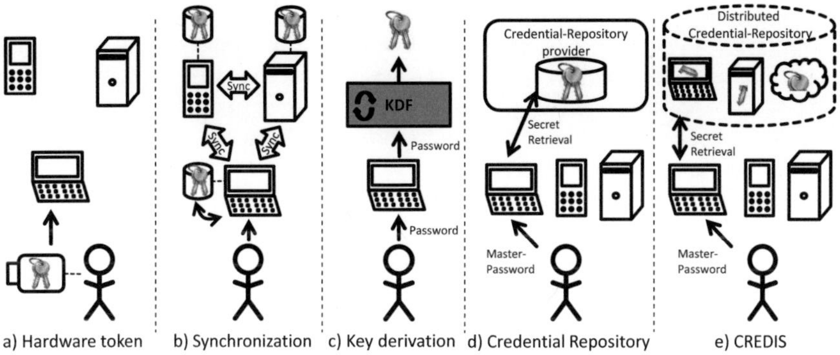

Figure 5.1: Approaches for credential management [KMH13].

In the following, we will address the problem of securely trading a (simple) human-generated password for a strong cryptographic secret with high entropy on arbitrary devices without storing permanently the secrets on those devices.

Key derivation functions (KDFs) [YY05, Kal00] were proposed to derive secure cryptographic secrets from weak passwords (see Figure 5.1c). As the image of a KDF can only be as big as the preimage, the entropy of the derived cryptographic secrets can only be as big as the password entropy. This leads to the problem of *offline attacks* on the password. For instance, to determine the cryptographic secret that can be used to decrypt the data, it suffices for the attacker to guess passwords (for instance based on a dictionary), use the KDF to derive the corresponding cryptographic secret, and check whether the password was correct by trying to decrypt the data with the derived secret.

KDFs address the problem of offline attacks by deliberately making the KDF inefficient to compute. Thus, an attacker has to spend more time on deriving the cryptographic secret from the guessed password. However, the KDF cannot be arbitrarily inefficient, as also users have to use it to derive their secrets. This trade-off constrains the applicability of KDFs for users that tend to use simple passwords as a study of leaked user passwords showed [WACS10]. With 50000 guesses an appropriately trained password cracking tool cracked more than 20% of the 8-character human-generated passwords that are contained in various lists of leaked passwords. Under the assumption that it takes 10 seconds to evaluate the KDF, it would only take around 6 days to make 50000 sequential password guesses. In fact, the threat is even worse as KDFs allow the attacker to parallelize the password guessing.

Another approach to trade weak passwords for strong secrets is to store the secrets in a **credential repository** that is hosted by a party that is trusted to enforce the confidentiality of the secrets and to only release secrets in exchange for the correct password (see Figure 5.1d). Once a user needs a secret on an end-device, the end-device can retrieve the secret from the credential repository by providing the user's weak password. After the secret is no longer needed on the end-device it can be discarded. Thus, if the end-device is compromised in the future, the user's secrets remain safe. Compared to KDFs, offline attacks on the password can be mitigated by letting the trusted party enforce lockout policies and allowing only a specific number of password guesses. For instance, such a lockout policy could be that the users account is locked for a certain time period after 3 incorrect passwords were sent to the credential repository. Thus, even very simple passwords can provide a high level of security. Compare this, for instance, to PIN numbers that are required to withdraw money from an ATM..

One drawback of credential repositories is that most users do not have a party that they trust unconditionally to enforce the confidentiality of their secrets. Even a server that is operated by the user at home can be prone to third party attacks. Another drawback is the fact that a credential repository constitutes a single-point-of-failure. Once the credential repository is unavailable, users do no longer have access to their secrets on any device.

5.2 Specific Research Question: Distributed Credential Repositories

Many users have many providers/devices (in the following: *stores*) at their disposal that could host credential repositories. However, in many cases the risk of storing secrets at any of those stores is perceived as too high. If the store is compromised, the secrets have to be assumed to be revealed to the attacker as well. Nevertheless, users are aware of the fact that the store can be compromised with a certain probability, but there is also the chance that the store remains uncompromised. For instance, a user typically acts on the assumption that her desktop computer is not compromised while at the same time the user is fully aware that there is a chance the computer is compromised by an attacker. Other examples for potentially compromised stores include smartphones, home-entertainment systems as well as external providers such as home organizations and cloud providers.

In this chapter, we aim to build distributed credential repositories based on multiple stores (see Figure 5.1e) that preserves the confidentiality of secrets with a sufficiently high probability. Distributed credential repositories split up each secret and store the fragments on different stores. This reduces the risk of secret disclosure since an attacker has to compromise all stores to restore the secret. However, all stores that store a fragment of a secret have to be available to retrieve the secret. In particular, if a secret is split up into n fragments which are stored on n stores, the secret is not available to the legitimate user if a single one of the n stores is unavailable. Thus, there exists a trade-off between secret availability and secret confidentiality. Which degree of secret confidentiality and secret availability is required depends on the deployment scenario. The research question we address in this chapter is:

How can the confidentiality of secrets be tuned in distributed credential repositories to satisfy deployment scenario specific confidentiality and availability requirements?

5.3 Related Work

Password Authenticated Key Exchange (PAKE) protocols [Jab96, BM93, BM92, STW95] allow users to securely establish an authenticated channel to a store based on a weak user password. Augmented PAKE (A-PAKE) protocols [GMR06, HK99] avoid storing password equivalent material at the store. Thus, if the store is compromised, the password and the stored secrets are not necessarily compromised. However, a compromised store will not enforce lockout policies anymore and therefore may perform password guessing attacks. As A-PAKE protocols only support a single store, the confidentiality of the secrets depends on whether the single store is compromised.

Secret sharing protocols [Bei11, Sha79, ISN89, vD95, BI93] split a secret on multiple parties and thus enhance confidentiality by not relying on the assumption that a single party is uncompromised. Secret sharing protocols enforce that the secret can only be restored if a specific subsets of stores collaborate. The set of the store subsets that are allowed to restore the secret forms the so-called *access structure* [Bei11]. For instance, an access structure can dictate that the secret may only be restored if S_1 and S_2, or S_2 and S_3 collaborate.

Traditional secret sharing protocols [Bei11, Sha79, ISN89, vD95, BI93] do not authenticate parties that want to restore the secret via password authentication[5]. Therefore, they cannot be applied to the problem setting of distributed credential repositories to enable trading weak passwords for strong secrets. The trade-off between availability and confidentiality also exists for secret sharing protocols and is determined by the enforced access structure. However, secret sharing protocols aim to enforce such defined access structure rather than tuning the access structure to deployment scenario requirements. The methodology of the Credis approach that is presented in this chapter to tune the trade-off between availability and confidentiality cannot only be applied to build distributed credential repositories but also to determine an access structure for traditional secret sharing schemes.

Threshold Password Authenticated Key Exchange (T-PAKE) protocols [MSJ02, FK00, Jab01] combine the authentication capabilities of PAKE protocols with the confidentiality enhancing concepts of traditional secret sharing schemes. T-PAKE protocols allow users to split up secrets on n stores. The user can retrieve the secrets based on a weak user-generated password. T-PAKE protocols enforce that 1) no store is able to reveal the secrets, 2) no store gets access to password equivalent data when authenticating the user, and 3) password guessing attacks are only possible if all n stores are compromised and collaborate. If at most $n - 1$ stores are compromised, the remaining honest stores mitigate password guessing attacks by enforcing a lockout policy that allows only a limited number of password guesses before requests are no longer answered. Ford et al. proposed such a protocol [FK00] which relies on previously authenticated channels to the stores. Jablon et al. [Jab01] extended the method of Ford et al. so that these previous authenticated channels are no longer necessary.

T-PAKE protocols can be used to distribute a secret on multiple stores. However, the protocols do not address the trade-off between availability and confidentiality.

5 In fact, authentication is beyond the scope of these protocols.

If a cryptographically strong secret is split up on n stores, all n stores have to be compromised to make password guessing attacks possible. However, all n stores have to be available to retrieve the secret. Credis uses the T-PAKE protocols as security mechanisms to outsource secrets and automatically adjusts the trade-off between availability and confidentiality to satisfy the deployment scenario requirements.

RAID concepts [PGK88] can be applied to enhance availability by redundantly storing data on multiple disks. Likewise, they can be used to increase the performance of read operations by fragmenting the stored data on multiple disks. Due to redundantly storing data, more disk space is required. This trade-off between availability, performance, and storage capacity in RAID systems is related to the trade-off between availability and confidentiality in distributed credential repositories. However, compared to the problem setting that is addressed by Credis, in RAID systems, disks (stores in the case of credential repositories) are assumed to have homogeneous properties which simplifies the resulting optimization problem a lot.

5.4 Distributed Credential Repositories: Credis

In this section we present Credis (**C**redential **RE**pository **D**istributed on **I**nhomogeneous **S**ystems), an approach that builds distributed credential repositories based on stores that are heterogeneous in terms of vulnerability as well as availability. Credis is tunable with regard to secret availability and confidentiality and, thus, can be used to tune the trade-off between those two characteristics. Credis allows the user to specify requirements with regard to secret availability and confidentiality that the distributed credential repository has to satisfy and automatically generates a strategy on how the stores at hand can be used to satisfy these requirements.

We introduce the general concept of Credis in Section 5.4.1 and show how the trade-off between secret availability and confidentiality can be tuned. In Section 5.4.2 we show how deployment scenario requirements can be specified by the user. In Section 5.4.3 we propose an automated method to fragment the secrets on stores in such a way that the deployment scenario requirements are satisfied and either secret availability or confidentiality is optimized. Furthermore, in Section 5.4.4 we provide efficient heuristics that can be used to satisfy the user's requirements in deployment scenarios with a large number of stores. We evaluate and discuss the Credis approach in Section 5.4.5 and conclude the chapter in Section 5.5.

5.4.1 General Concept

Credis addresses the setting that is shown in Figure 5.1e). A user needs a strong secret on an end-device. To retrieve the secret, the user enters a potentially weak password at the end-device. The end-device uses this password to retrieve the strong secret from a number of stores that form a distributed credential repository.

Attacker model: Credis assumes that the end-device is not compromised by the attacker while the password is entered and the retrieved secrets are cached. The attacker is able to compromise each store with a certain probability and strives to reveal the user's secrets or the password that can be used to retrieve those secrets.

In the following we assume without loss of generality that only a single secret will be stored in the credential repository[6]. We denote this secret as *key* in the following.

Credis relies on the existing T-PAKE protocols that were introduced in Section 5.3 to split the key in multiple *fragments* which are stored on distinct stores. The protocols guarantee that the stores release the fragments only if the correct password is entered. Furthermore, during the key retrieval process, the protocols ensure that the password cannot be revealed by an attacker that compromised a subset of the stores that store the fragments.

> *Definition - Key-availability:* Key-availability denotes the probability that the key is available to a legitimate user at a given point in time. The key is available if all fragments of at least one fragment bundle are available.

> *Definition - Key-vulnerability:* Key-vulnerability denotes the probability that the key can be viewed by an attacker between the point in time when the key is stored in the distributed credential repository and the time it is deleted. The confidentiality of the key is undermined if all fragments of at least one fragment bundle are accessible by the attacker.

One approach to store the key on the stores is to put one fragment at each store. This minimizes the vulnerability of the credential repository (*key-vulnerability*) as an attacker would have to compromise all stores to undermine the confidentiality of the stored key. However, the credential repository would only be available (*key-availability*) if all stores are available. Thus, there is a trade-off between key-vulnerability and key-availability. To allow for a fine-grained tuning of this trade-off, Credis splits the key not only once but multiple times, forming *fragment bundles*. The key can be reconstructed from any fragment bundle. Fragments of different bundles are not compatible, i.e., all fragments of at one least fragment bundle need to be known to re-construct the key. The fragments of the bundles can be distributed on the stores independently from one another. The distribution of fragments on stores is denoted as *key fragment distribution* in the following. The key-availability and the key-vulnerability of the distributed credential repository directly depend on the key fragment distribution.

In the following, the notation $B_1 = \{S_1, S_2, \ldots\}$ expresses that one fragment of the bundle B_1 resides on each store S_i.

> *Example:* An exemplary key fragment distribution is shown in Figure 5.2. If the stores S_3 and S_4 are not available, the key can still be retrieved by retrieving fragment bundle B_1 from the stores S_1 and S_2. However, for an attacker it suffices to compromise S_1 and S_2 to compromise the secret.

6 In fact this single key can be used to encrypt an arbitrary number of other secrets. Given that the utilized encryption scheme is secure, the resulting ciphertexts can be distributed to all stores without harming the confidentiality of the secrets.

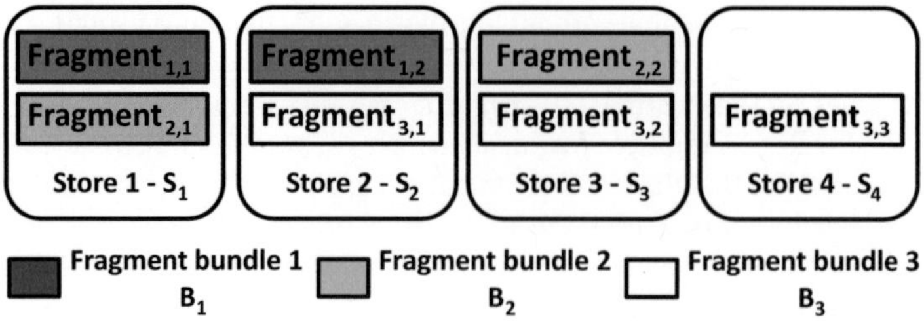

Figure 5.2: Exemplary key fragment distribution [KMH13].

How the trade-off between key-vulnerability and key-availability should be tuned depends on the user's requirements. While some applications have higher requirements concerning the availability of the key, others prefer a small vulnerability over a high availability of the key. We show in Section 5.4.2 how the user can specify the scenario and its requirements. In Section 5.4.3 we show that the problem of finding a key fragment distribution that optimally satisfies the deployment scenario requirements is NP-hard and propose a model to formalize the optimization problem. We show how the problem can be optimally solved in Section 5.4.3 and propose heuristics to solve large problem instances with acceptable performance in Section 5.4.4.

5.4.2 Scenario Definition

A scenario contains a set of stores with individual properties and deployment scenario requirements regarding the key-vulnerability and key-availability.

Store Properties

Each store at hand can have individual properties with regard to availability and vulnerability.

> *Definition - Store-availability:* Store-availability of a store denotes the probability that the store is available at a given point in time.

> *Definition - Store-vulnerability:* Store-vulnerability denotes the probability that the store is compromised by an attacker at any point between the time of the key deployment in the distributed credential repository and the time the key is deleted or replaced by a new key.

In the following, we denote the store-vulnerability of a store S_i as $V(S_i)$ and its store-availability as $A(S_i)$. For each store, Credis expects $V(S_i)$ and $A(S_i)$ as input parameters. A discussion on how to find meaningful values for store-availability and store-vulnerability is provided in Section 5.4.5. For simplicity, we make the assumption that both the store-availability and the store-vulnerability of one store are independent from those of another store. We discuss this assumption in Section 5.4.5.

Deployment scenario Requirements

Credis determines a key fragment distribution that satisfies deployment scenario requirements with regard to key-vulnerability and key-availability. These requirements have to be specified by the user of Credis. Furthermore, Credis allows the user to specify which key fragment distribution that satisfies the requirements is considered optimal by an optimization criterion.

Credis allows the user to specify the following **boundary conditions** to define feasible key fragment distributions:

- **Maximum key-vulnerability**: Specifies the upper acceptable bound for key-vulnerability, i.e., the probability of the key becoming compromised.

- **Minimum key-availability**: Specifies the lower acceptable bound for key-availability, i.e., the probability of the key being available.

- **Maximum number of bundles**: "Specifies the upper acceptable bound for the number of bundles to prevent online dictionary attacks. Depending on the lockout policy of the stores, the number of bundles directly affects the number of possible password guesses an attacker can make before being blacklisted by the stores (e.g., with 3 bundles and the lockout policy of only tolerating 5 failed attempts to enter the password correctly, an attacker can make 15 guesses on the password before being locked out)." [KMH13]

Credis allows the user to specify the following **optimization criteria** to specify which key fragment distribution is considered optimal:

- **Key-availability maximization**: "Requests the feasible key fragment distribution with maximum probability of the key being available." [KMH13]

- **Key-vulnerability minimization**: "Requests the feasible key fragment distribution with minimum probability of the key becoming compromised." [KMH13]

- **Joint optimization:** "Key availability and key-vulnerability can be weighted to optimize for a ratio of the two properties (e.g., 1*key-availability - 5*key-vulnerability)." [KMH13]

We denote a key fragment distribution that is feasible with regard to the boundary conditions and optimal with regard to the optimization criterion as *optimal key fragment distribution* in the following. Finding such an optimal key fragment distribution for a given scenario constitutes an optimization problem for which we provide solving methods in the following section.

5.4.3 Finding Optimal Solutions

In this section we show that finding an optimal key fragment distribution is NP-hard and provide a linearization of the optimization problem that enables us to model it as a common *Integer Linear Programming* (ILP) problem. This problem can be solved by existing ILP solvers. We evaluate the performance and scalability of this approach in Section 5.4.5.

NP-Hardness Proof

We originally published the following proof in [KMH13].

Proof. To prove the NP-hardness of finding an optimal key fragment distribution, the NP-hard *KNAPSACK* problem can be reduced to finding an optimal key fragment distribution for a specific scenario:

The KNAPSACK Problem:

Given: A *KNAPSACK* problem instance. A set of items $I = \{i_1, i_2, \ldots, i_n\}$ with non-negative values $c_1, \ldots, c_n$ and non-negative weights $w_1, \ldots, w_n$ that are assigned to each item is given. Furthermore, a weight capacity W of the knapsack exists.
Wanted: A subset of items $K \subseteq I$ so that $\sum_{i_j \in K} w_j \leq W$ holds true and $V = \sum_{i_j \in K} c_j$ is maximized.

Reduction:

1. Define the deployment scenario as follows:
   - Stores: $\mathcal{S} = \{S_1, S_2, \ldots, S_n\}$
   - Minimum key-availability boundary condition: A
   - Maximal number of fragment bundles: 1
   - Optimization criterion: Key-vulnerability minimization

   Choose the minimal key-availability as $A = 2^{-W}$, store-availability of each S_j as $a_j = 2^{-w_j}$, and the store-vulnerability as $v_j = 2^{-c_j}$.
2. Find the optimal key fragment distribution for the deployment scenario. This will result in one fragment bundle B_1.
3. Solution transformation: For each S_j in the fragment bundle B_1 put item i_j in the knapsack.

This reduction is achieved on polynomial time. To prove the correctness of the reduction, we show that the solutions for any *KNAPSACK* problem instance that are determined this way are both feasible and optimal.

Feasibility: As according to the deployment scenario requirements, B_1 is the only fragment bundle, the availability of the key can be calculated as $\prod_{S_j \in B_1} a_j$. Due to the minimum key-availability boundary condition policy it holds that:

$$\begin{aligned} \prod_{S_j \in B_1} a_j \geq & \quad A \\ \Leftrightarrow \prod_{i_j \in K} 2^{-w_j} \geq & \quad 2^{-W} \\ \Leftrightarrow \sum_{i_j \in K} w_j \leq & \quad W \end{aligned}$$

Thus, the weight of all items in the knapsack does not exceed the knapsacks capacity. Consequently, each feasible bundle choice can be mapped on a feasible subset of items in the knapsack and vice versa.

Optimality: According to the deployment scenario requirements, B_1 is the only fragment bundle. Thus, the key-vulnerability, i.e., the probability that the key is compromised, can be calculated as $\prod_{S_j \in B_1} v_j$. As the key fragment distribution is optimal in the sense that key-vulnerability is minimized, it holds true that the vulnerability of other feasible fragment bundle choices ($\mathcal{B}^{feas}$) is bigger or equal to that of B_1:

$$
\begin{aligned}
& \forall B' \subseteq \mathcal{B}^{feas} : \prod_{S_j \in B'} v_j \geq \prod_{S_j \in B_1} v_j \\
\overset{Correctness}{\Leftrightarrow} & \forall K' \subseteq \mathcal{K}^{feas} : \prod_{i_j \in K'} 2^{-c_j} \geq \prod_{i_j \in K} 2^{-c_j} \\
\Leftrightarrow & \forall K' \subseteq \mathcal{K}^{feas} : \sum_{i_j \in K'} c_j \leq \sum_{i_j \in K} c_j
\end{aligned}
$$

Thus, as all other valid item choices ($\mathcal{K}^{feas}$) that fit in the knapsack are less or equally valuable, K is the optimal solution of the *KNAPSACK* problem instance. □

Mapping on an ILP problem

Parts of the section on mapping the problem of finding the optimal key fragment distribution on an ILP problem are quoted verbatim from our original publication of Credis [KMH13]. To map the problem of finding the optimal key fragment distribution on an ILP problem, we need to be able to determine the key-availability and the key-vulnerability of given key fragment distributions. Bundles can overlap in the sense that one or more stores store fragments of each bundle. Therefore, the stochastic event of a bundle being compromised is not independent of the stochastic event of another bundle being compromised if both bundles overlap.

> *Example:* Let us consider an exemplary key fragment distribution consisting of two bundles $B_1 = \{S_1, S_2\}$ and $B_2 = \{S_2, S_3\}$ as a running example in this section with store-vulnerabilities of V(S_1)=0.1, V(S_2)=0.2, V(S_3)=0.15 and store-availabilities A(S_1)=0.7, A(S_2)=0.8, A(S_3)=0.9. While it holds that $V(B_1) = V(S_1) * V(S_2) = 0.02$ and $V(B_2) = V(S_2) * V(S_3) = 0.015$, the probability of the key becoming compromised (i.e., at least one bundle becoming compromised) does not amount to $1 - (1 - V(B_1)) * (1 - V(B_2)) = 0.0494$ as the vulnerabilities $V(B_1)$ and $V(B_2)$ both depend on the vulnerability of S_1.

In the following we will denote stochastic events as *relevant*, if their occurrence implies a compromised or available bundle and, thus, key.

> *Example:* The bundle $B_1 = \{S_1, S_2\}$ makes both events, "S_1, S_2 and S_3 being compromised" as well as "S_1, S_2 being compromised and S_3 not being compromised" relevant, as in both cases all fragments of B_1 are compromised. Concerning key-availability, B_1 makes the events of "S_1, S_2 and S_3 being available" as well as "S_1, S_2 being available and S_3 being not available" relevant, as in both cases the fragments of B_1 can be retrieved.

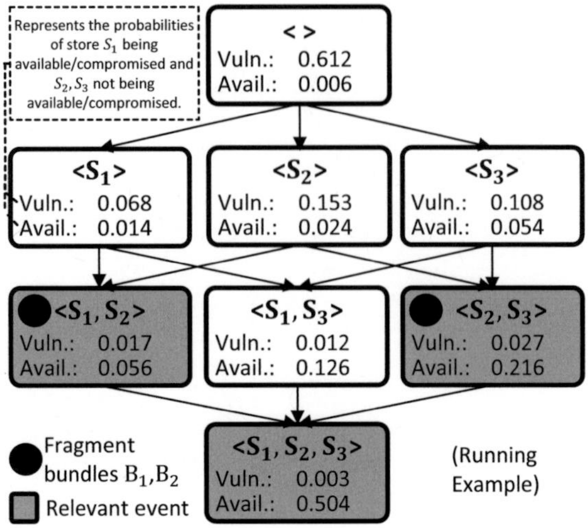

Figure 5.3: Examplary event relevance graph based on the stores of the running example: $V(S_1)$=0.1, $V(S_2)$=0.2, $V(S_3)$=0.15, $A(S_1)$=0.7, $A(S_2)$=0.8, $A(S_3)$=0.9. [KMH13].

The relevant events that concern key-availability are exactly those events in which the key is available, the relevant events that concern key-vulnerability are exactly those events in which the key is compromised. Thus, the sum of the occurrence probabilities of relevant events concerning key-availability equals the key-availability of the key fragment distribution. Analogously, the sum of the occurrence probabilities of relevant events concerning the key-vulnerability equals the key-vulnerability.

To make the problem of rating key fragment distributions more explicit, we model it as a graph that we denote as *event relevance graph* in the following. The event relevance graph for our exemplary key fragment distribution is depicted in Figure 5.3. Event relevance graphs can be interpreted as follows: Each node $< S_{i_1}, S_{i_2}, \ldots >$ can be interpreted as the event of the stores $S_{i_1}, S_{i_2}, \ldots$ being available and all remaining stores not being available. The occurrence probability of these events is denoted by "Avail." in Figure 5.3. Each node can also be interpreted as the event of the stores $S_{i_1}, S_{i_2}, \ldots$ being compromised and all remaining stores not being compromised. The occurrence probability of these events is denoted by "Vuln." in Figure 5.3.

> *Example:* In Figure 5.3, node $< S_1 >$ corresponds to the event of S_1 being compromised while S_2 and S_3 are not compromised (occurrence probability: $V(S_1) * (1 - V(S_2)) * (1 - V(S_3)) = 0.068$). Node $< S_1 >$ also corresponds to the event of S_1 being available while S_2 and S_3 are not available (occurrence probability: $A(S_1) * (1 - A(S_2)) * (1 - A(S_3)) = 0.014$).

Each directed edge from an event e_1 to an event e_2 in the graph expresses that if a bundle is compromised/available in case of occurrence of the event e_1 it would also be compromised/available if event e_2 occurs. In particular, these semantics of the edges imply that, if an event is relevant, all events that can be reached from this event in the graph have to be relevant, too.

> *Example:* A bundle that is compromised if S_1 is compromised and S_2, S_3 are not compromised (node $< S_1 >$), it would also be compromised if S_1 and S_2 are compromised and S_3 is not compromised (node $< S_1, S_2 >$). Analogously, a bundle that is available if S_1 is available and S_2, S_3 are not available (node $< S_1 >$), it would also be available if S_1 and S_2 are available and S_3 is not available (node $< S_1, S_2 >$). Thus, the event relevance graph contains an edge from node $< S_1 >$ to node $< S_1, S_2 >$. Our exemplary key fragment distribution that consists of the bundles $B_1 = \{S_1, S_2\}$ and $B_2 = \{S_2, S_3\}$ is expressed in Figure 5.3 by black circles. Node $< S_1, S_2 >$ is relevant, as B_1 would be compromised/available in the event of S_1 and S_2 being compromised/available and S_3 not being compromised/available. The same applies for node $< S_2, S_3 >$ and B_2. Furthermore, all nodes that can be reached from $< S_1, S_2 >$ and $< S_2, S_3 >$ are relevant as well due to the semantics of the edges.

Rating key fragment distributions: Both the probability of at least one of the bundles being compromised and at least one of the bundles being available can be calculated by summing up the "Vuln." and "Avail." occurrence probabilities of the relevant nodes respectively. Thus, the problem of finding an optimal key fragment distribution is reduced to finding a choice of bundles so that both the sum of the "Avail." property and the sum of the "Vuln." property of the relevant nodes satisfy the deployment scenario requirements.

> *Example:* The key-availability of the key fragment distribution shown in Figure 5.3 amounts to $0.056 + 0.504 + 0.216 = 0.776$. The key-vulnerability amounts to $0.017 + 0.003 + 0.027 = 0.047$.

This resulting problem is a linear problem and can be formulated as the following ILP problem:

Constants:

N	: Set of nodes in the event relevance graph.
V	: Maximum key-vulnerability deployment scenario requirement.
V_i	: Vulnerability of node n_i in the event relevance graph.
A	: Minimum key-availability deployment scenario requirement.
A_i	: Availability of node n_i in the event relevance graph.
$maxBdls$	: Maximum number of key fragment bundles.
$E(n)$	: Set of (child) nodes that are reachable from node n.
$I(n)$	: Set of (parent) nodes that can reach node n.

Free variables:

x_i	: Node n_i is relevant.
y_i	: Node n_i represents a bundle.

ILP optimization problem:

$$\max \sum_{n_i \in N} A_i * x_i \textbf{ or } \min \sum_{n_i \in N} V_i * x_i \quad (5.1)$$

s.t.

$$\sum_{n_i \in N} V_i * x_i \leq V \quad (5.2)$$

$$\sum_{n_i \in N} A_i * x_i \geq A \quad (5.3)$$

$$\forall n_j \in N : \frac{1}{|E(n_j)|} * \sum_{n_i \in E(n_j)} x_i \geq x_j \quad (5.4)$$

$$\forall n_j \in N : x_j \geq y_j \quad (5.5)$$

$$\forall n_j \in N : \sum_{n_i \in I(n_j)} x_i + y_j \geq x_j \quad (5.6)$$

$$\forall n_j \in N : 1 - \frac{\sum_{n_i \in I(n_j)} x_i}{|I(n_j)|} \geq y_j \quad (5.7)$$

$$\sum_{n_i \in N} y_i \leq maxBdls \quad (5.8)$$

$$\forall n_j \in N : x_i \in \{0,1\} \quad (5.9)$$

$$\forall n_j \in N : y_i \in \{0,1\} \quad (5.10)$$

The target function can be defined depending on the deployment scenario requirements' optimization criterion. In the listed ILP problem, target functions for maximizing the key-availability or minimizing the key-vulnerability are shown in equation (5.1). However, arbitrary (linear) target functions can be specified (e.g., for joint optimization of key-availability and key-vulnerability).

Constraint 5.2 and Constraint 5.3 limit the key-vulnerability and the key-availability of the feasible solutions according to the deployment scenario requirements. Constraint 5.4 assures, that nodes can only be marked as relevant, if all their child nodes are relevant. Each node that represents a bundle needs to be relevant (Constraint 5.5). Furthermore, each relevant node has to represent a bundle or a parent node has to be relevant (Constraint 5.6). Constraint 5.7 is optional and enforces that a node may only represent a bundle if no parent node is relevant. Without Constraint 5.7, the solution of the ILP problem can contain redundant, i.e., bundles that contain all stores of another bundle. Such bundles can be purged from the solution since they neither affect key-availability nor key-vulnerability. The number of bundles has to be limited according to the specified deployment scenario requirements (Constraint 5.8). Furthermore, nodes can either be relevant ($x_i = 1$) or not relevant ($x_i = 0$) (Constraint 5.9). The same has to hold for nodes that represent a bundle (Constraint 5.10).

Solving the ILP problem yields the optimal choice of bundles. Solving ILP problems is NP-hard and the number of nodes that have to be taken into account depends exponentially on the number of stores ($N = 2^{|stores|}$). However, we will show in Section 5.4.5 that the approach performs well for problems with up to 9 stores.

5.4.4 Finding Heuristic Solutions

While a problem size of 9 stores seems to be enough for common scenarios today, the advent of the internet of things [AIM10] makes scenarios with more stores seem realistic. To enable the application of Credis in scenarios with a large number of stores, we propose efficient heuristics that approximate the optimal solution for deployment scenarios in which key-availability has to be maximized (*MaxAvail*) and for deployment scenarios in which key-vulnerability has to be minimized (*MinVuln*). Parts of the sections that introduce the MaxAvail and MinVuln heuristics are quoted verbatim from our original publication of Credis [KMH13].

MaxAvail Heuristic

The pseudocode algorithm of the **MaxAvail** greedy heuristic is listed in Algorithm 1. The basic principle consists of adding stores to a bundle B_{cur} until the maximum key-vulnerability condition is satisfied (cf. line 7). The choice of which store to add in each step depends on the benefit for adding the store (cf. line 8-19). The store's benefit is calculated by dividing the gain in key-availability Δa by the gain in key-vulnerability Δv that would result from adding the current bundle including the store to the solution. The store that benefits the key fragment distribution the most is added to the current bundle B_{cur} (cf. line 23-24). This step of rating stores and adding them to B_{cur} is repeated until the maximum key-vulnerability condition is satisfied by the key fragment distribution. Once the maximum key-vulnerability condition is satisfied, the bundle B_{cur} is committed to the solution (cf. line 26-27) and the process is repeated for a new bundle. This is done until adding additional bundles does not yield any availability gain (cf. line 20-21) or the maximum number of bundles as specified in the deployment scenario requirements is reached (cf. line 4).

> *Example:* An exemplary run of the MaxAvail heuristic on the example introduced in Section 5.4.3 is shown in Figure 5.4 for a maximum key-vulnerability boundary condition of 0.05. The algorithm starts with an empty fragment bundle set that does not induce any vulnerability ($vuln_{total}$ = 0). In the first step, the benefit of adding each store to an empty bundle is calculated by dividing the gain in key-availability by the gain in key-vulnerability. For instance, by adding store S_1 to the bundle, the events of $< S_1 >, < S_1, S_2 >, < S_1, S_3 >, < S_1, S_2, S_3 >$ (cf. Figure 5.3) become relevant. Their summed up key-vulnerability values amount to 0.1 and their summed up key-availability values amount to 0.7. Thus, the benefit of adding S_1 is 7. As the benefits of adding S_2 or S_3 are less than that, S_1 is added to the bundle. The key-vulnerability induced by the bundle ($vuln_{cur}$ = 0.1) is bigger than the maximum key-vulnerability 0.05. Thus, another store has to be added to the current bundle (cf. line 7 in Algorithm 1). The process of rating the benefits of the remaining stores is repeated and the best one (S_3) is added to the bundle. As the key-vulnerability ($vuln_{cur}$ = 0.015) that is induced by the bundle is less than the maximum key-vulnerability, the bundle $\{S_1, S_3\}$ can be added to the fragment bundle set.

Algorithm 1 MaxAvail heuristic

```
procedure MaxAvail(stores, vuln_max, maxBdls)
    bundles ← ∅
    vuln_total ← 0
    while |bundles| ≤ maxBdls do
        B_cur ← ∅
        vuln_cur ← 1
        while vuln_total + vuln_cur > vuln_max do
            benefit_best ← −1
            for S ∈ stores do
                Δv, Δa ← Rate(B_cur ∪ {S}, bundles)
                if Δv ≠ 0 then
                    benefit_cur ← Δa/Δv
                    if benefit_cur > benefit_best then
                        benefit_best ← benefit_cur
                        bestS ← S
                        bestVLoss ← Δv
                    end if
                end if
            end for
            if benefit_best = −1 then
                return bundles
            end if
            B_cur ← B_cur ∪ {bestS}
            vuln_cur ← bestVLoss
        end while
        bundles ← bundles ∪ {B_cur}
        vuln_total ← vuln_total + vuln_cur
    end while
    return bundles
end procedure
```

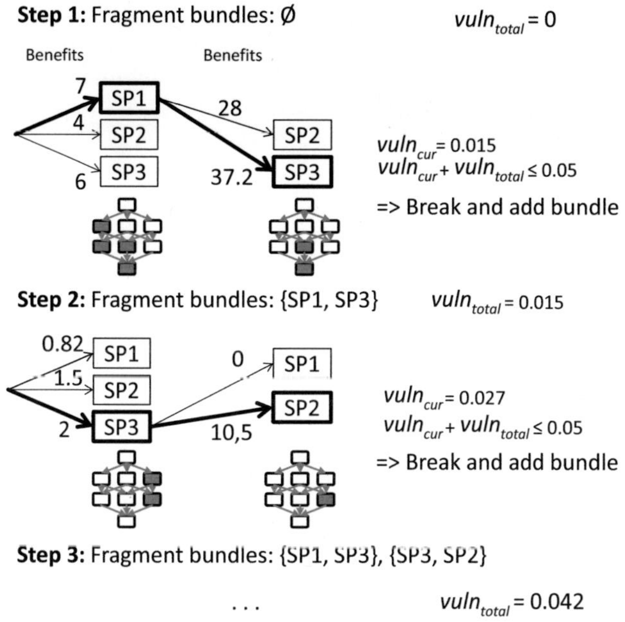

Figure 5.4: Exemplary run of the MaxAvail heuristic [KMH13].

After the first bundle is submitted to the fragment bundle set, the algorithm checks whether additional bundles can be added to improve the solution (cf. Step 2 in Figure 5.4). The heuristic starts with an empty bundle and rates the benefit of adding each store, taking the already submitted bundles into account: For instance, by adding store S_3 to the bundle, only the events of $< S_3 >$ and $< S_2, S_3 >$ additionally become relevant, as $< S_1, S_3 >$ and $< S_1, S_2, S_3 >$ are already relevant due to the submitted bundle of Step 1. Again, stores are added until the sum of the induced key-vulnerability of the submitted bundles ($vuln_{total}$) and of the current bundle ($vuln_{cur}$) is less than the maximum key-vulnerability. If this is not possible, the heuristic terminates and returns the submitted bundles as solution. The procedure of creating bundles and adding them to the fragment bundle set is repeated until the maximum number of bundles as specified by the user is reached and the fragment bundle set is returned as the solution.

The pseudocode procedure makes use of the *Rate(...)* procedure that is listed in Algorithm 2 to determine the gain in availability Δa and the gain in vulnerability Δv that would result from adding a bundle B_{cur} to the bundles that are already contained in the solution. The steps the algorithm performs are exemplarily shown in Figure 5.5. The depicted graphs can be interpreted as described in Section 5.4.3. In Figure 5.5 the events that are relevant due to B_{cur} and the events that are relevant due to other bundles B_1 and B_2 that are already contained in the solution are shown. When adding B_{cur} to the solution, one event will be relevant additionally and influences both key-availability and key-vulnerability. However, the naive approach of keeping track of all relevant events and calculating the influence of B_{cur} by only considering the *additionally* relevant events does not scale well, due to the fact that $2^{|stores|}$ nodes would have to be tracked.

Algorithm 2 Rate algorithm

1: **procedure** Rate(B_{cur}, $bundles$)
2: $\Delta v \leftarrow 0$
3: $\Delta a \leftarrow 0$
4: **for all** $bundleCombination \in 2^{bundles}$ **do**
5: $curIntersection \leftarrow B_{cur}$
6: **for all** $B \in bundleCombination$ **do**
7: $curIntersection \leftarrow curIntersection \bigcap B$
8: **end for**
9: **if** $|bundleCombination|$ is uneven **then**
10: $\Delta v \leftarrow \Delta v - \prod_{S \in curIntersection} V(S)$
11: $\Delta a \leftarrow \Delta a - \prod_{S \in curIntersection} A(S)$
12: **else**
13: $\Delta v \leftarrow \Delta v + \prod_{S \in curIntersection} V(S)$
14: $\Delta a \leftarrow \Delta a + \prod_{S \in curIntersection} A(S)$
15: **end if**
16: **end for**
17: **return** $\Delta v, \Delta a$
18: **end procedure**

To work around that, our algorithm initially naively determines the vulnerability of B_{cur} by calculating the product of the store-vulnerabilities of all stores in B_{cur}[7]. Probabilistic theory dictates that this product equals the sum of the events that are relevant because of B_{cur} (cf. Figure 5.5, Step 1). In order to correct the error of including events that are already relevant because of existing bundles, our algorithm subtracts the occurrence probabilities of the events in the intersection of B_{cur} and each existing bundle (line 10-11 in Algorithm 2). For instance, the summed up probabilities of the events in the intersection of B_{cur} and B_1 can be easily determined by calculating the vulnerability of the imaginary bundle $B_{cur} \bigcap B_1$ (cf. Figure 5.5, Step 2). By subtracting both the summed up probabilities of $B_{cur} \bigcap B_1$ and $B_{cur} \bigcap B_2$, another error is induced, as the event contained in both $B_{cur} \bigcap B_1$ and $B_{cur} \bigcap B_2$ is subtracted two times. To remedy that, the summed up probabilities of the events in the intersection of B_{cur}, B_1, and B_2 (cf. Figure 5.5, Step 3) are computed as explained and re-added to the result (line 13-14 in Algorithm 2).

7 For the sake of simplicity, we only address vulnerability in the explanation. However, all steps can be analogously performed for availability.

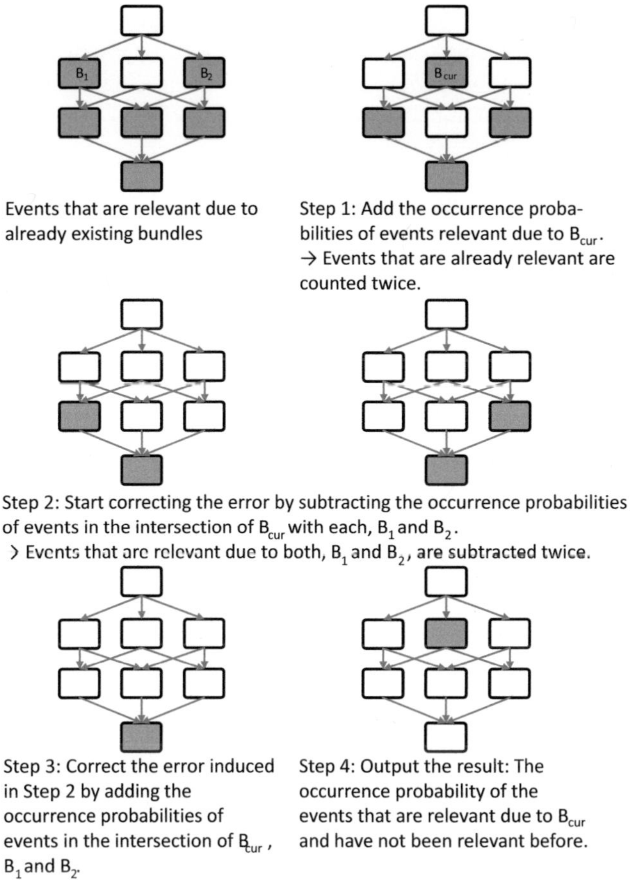

Figure 5.5: Exemplary run of the *Rate(...)* algorithm to rate the fragment bundle B_{cur} in the presence of the bundles B_1 and B_2 [KMH13].

The asymptotic execution time of the MaxAvail algorithm amounts to $O(m*n^2*2^m)$, where n constitutes the number of stores and m the maximum number of bundles as specified in the deployment scenario requirements. Therefore, it does not depend exponentially on the number of stores but only on the number of maximum bundles as specified in the deployment scenario requirements. We argue, that in realistic scenarios, this parameter will not be chosen exceptionally large to prevent online dictionary attacks that are performed by external attackers.

MinVuln Heuristic

The **MinVuln** binary search heuristic to solve key-vulnerability minimization problems can be constructed based on the MaxAvail heuristic. The key fragment distribution with the required key-availability and a minimized key-vulnerability is approximated by applying the MaxAvail heuristic multiple times to zone in on the minimally achievable key-vulnerability values. A target precision has to be specified that describes the tolerable absolute deviation from the minimum key-vulnerability. The pseudocode of MinVuln is shown in Algorithm 3.

Algorithm 3 MinVuln heuristic

```
1: procedure MinVuln(stores, avail_min, maxBdls, prec)
2:     vuln_max ← 0.5
3:     curPrec ← 0.5
4:     while curPrec > prec do
5:         curSol = MaxAvail(stores, vuln_max, maxBdls)
6:         curPrec ← curPrec / 2
7:         if avail_min ≤ Availability(curSol) then
8:             sol ← curSol
9:             vuln_max ← vuln_max − curPrec
10:        else
11:            vuln_max ← vuln_max + curPrec
12:        end if
13:    end while
14:    return sol
15: end procedure
```

The algorithm starts by using the MaxAvail heuristic to solve a key-availability maximization problem that has been constructed based on the *stores* and the maximum number of bundles (*maxBdls*) of the actual deployment scenario requirements (line 5). The maximum key-vulnerability boundary condition is initially set to 0.5 ($vuln_{max}$, line 2). If the key-availability of the solution satisfies the key-availability boundary condition of the original problem, the $vuln_{max}$ parameter is tightened by subtracting the current precision parameter ($curPrec$, line 7-9) to check if a feasible solution with a lower key-vulnerability exists. Otherwise $vuln_{max}$ is loosened by adding $curPrec$ (line 10-12) to find a feasible solution. Afterward, another key-maximization problem with the new $vuln_{max}$ boundary condition parameter is solved and the precision parameter is halved (line 6). This process is repeated until $curPrec$ falls below the required precision $prec$ (line 4). Once the algorithm terminates, the last key fragment distribution that adhered to the key-availability boundary condition of the original problem is returned (lines 8 and 14).

> *Example*: An exemplary run of the MinVuln heuristic is shown in Figure 5.6 for a scenario with a minimum key-availability requirement ($avail_{min}$) of 0.85 and a target precision of $\frac{1}{30}$. The precision variable $curPrec$ is initially set to 0.5. In the first step, the minimum key-availability requirement is substituted by a maximum key-vulnerability requirement of 0.5 ($vuln_{max}$). Solving this problem using the MaxAvail heuristic results in a solution with a key-availability of 0.99 which is higher than the

requested availability $avail_{min}$. In Step 2, the precision variable $curPrec$ is halved and to find potential solutions with a smaller key-vulnerability, $vuln_{max}$ is tightened by subtracting $curPrec$. Solving this problem results in a solution with a key-availability of 0.95. Thus, $(vuln_{max})$ is tightened again in Step 3. The solution of Step 3, however, results in a key-availability of 0.84 which violates the minimum key-availability requirement of the original scenario. Thus, $vuln_{max}$ is loosened by adding $curPrec$ in Step 4. This procedure is repeated until the precision variable is below $\frac{1}{30}$, the threshold defined by the passed $prec$ parameter. Once this happens, the last feasible solution is returned as a result. This is the solution of Step 4 in this example.

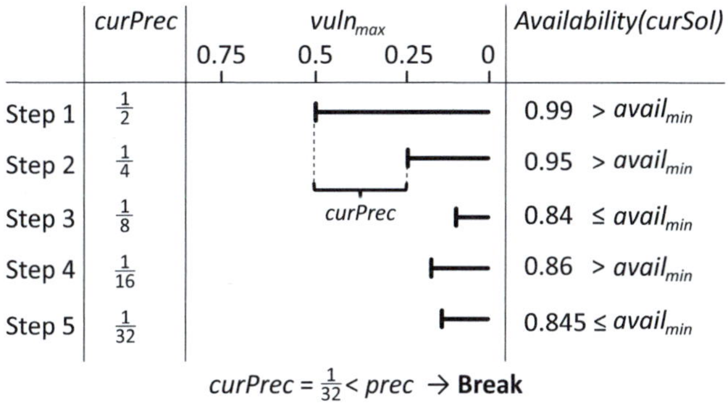

Figure 5.6: Exemplary run of the MinVuln heuristic ($avail_{min}$ = 0.85, $prec = \frac{1}{30}$) [KMH13].

The asymptotic execution time of MinVuln amounts to $O(log_2(\frac{1}{p}) * m * n^2 * 2^m)$, where p constitutes the precision parameter, n constitutes the number of stores and m the maximum number of bundles defined in the deployment scenario requirements. A more detailed performance evaluation of the heuristic will be presented in Section 5.4.5.

5.4.5 Evaluation and Discussion

Impact of the Store Properties on Availability and Vulnerability

To investigate the impact of the stores' vulnerability and availability properties on the properties of the key fragment distribution that is determined by Credis, we applied Credis for multiple deployment scenario requirements, each with a different set of stores that have heterogeneous properties. We investigated deployment scenarios with a minimum key-availability requirement and minimized key-vulnerability. The experimental setup and the key-vulnerabilities of the key fragment distributions that were determined by Credis are shown in Table 5.1. The stores that are contained in a scenario are marked with an "x".

The results show that even stores with a comparatively high store-vulnerability can be leveraged by Credis to achieve key fragment distributions that have a significantly lower key-vulnerability. For instance, in scenario 4, a key-vulnerability can be achieved that is less than $\frac{1}{5}$ of the key-vulnerability of scenario 6. Furthermore, the results show that relaxing the availability requirement by 4% can result in a key-vulnerability that is improved by orders of magnitude. For instance, comparing scenario 1 and 2, by relaxing the key-availability requirement by 4%, the key-vulnerability can be reduced to $\frac{1}{5}$ of the key-vulnerability of scenario 1.

Benefit of Tuning Availability and Vulnerability

To compare the Credis approach that allows to tune key-availability in exchange for key-vulnerability and vice versa, we compare it with three static approaches: storing the key on a single store (*single system*), replicating the key on all stores of the scenario (*replicated*) and fragmenting the key on all stores of the scenario (*partitioned*). The scenarios that we investigated contained a varying number or stores with a fixed store-vulnerability of 0.1 and store availability of 0.85. The key-vulnerabilities and key-availabilities of the solutions provided by the approaches for different numbers of stores are shown in Figure 5.7. The static approaches produce exactly one solution for each number of stores (see the single dots labeled with the number of stores of the scenario in Figure 5.7). Credis produces various solutions for each number of stores based on the boundary conditions, e.g., the minimum required key-availability (see the connected dots in Figure 5.7).

"While the number of stores does not affect the solution if the key is only stored on a single store (*single system* in Figure 5.7), it is affected when partitioning or replicating the key on all stores at hand. In case of partitioning the key, the key-vulnerability *and* the key-availability decrease by orders of magnitude when more stores are used (*partitioned* in Figure 5.7). In case of replicating the key, both key-availability and key-vulnerability increase by orders of magnitude (*replicated* in Figure 5.7). Thus, the *replicated* and the *partitioned* strategy are both extreme strategies. In fact their solutions constitute upper and lower bounds for achievable key-availability and key-vulnerability values, respectively." [KMH13]

	S_1	S_2	S_3	S_4	S_5	S_6	Min. Avail.	Vuln.
Vuln.	0.25	0.25	0.05	0.1	0.2	0.05		
Avail.	0.7	0.7	0.95	0.9	0.9999	0.99		
Scen.1	x	x	x	x	x	x	0.99	0.0016875
Scen.2	x	x	x	x	x	x	0.95	0.0003063
Scen.3	x	x	x	x	x		0.99	0.0183125
Scen.4	x	x	x	x	x		0.95	0.0061250
Scen.5	x	x	x	x			0.99	0.0915625
Scen.6	x	x	x	x			0.95	0.0306250

Table 5.1: Vulnerability of solutions for different scenarios with a key-availability boundary condition [KMH13].

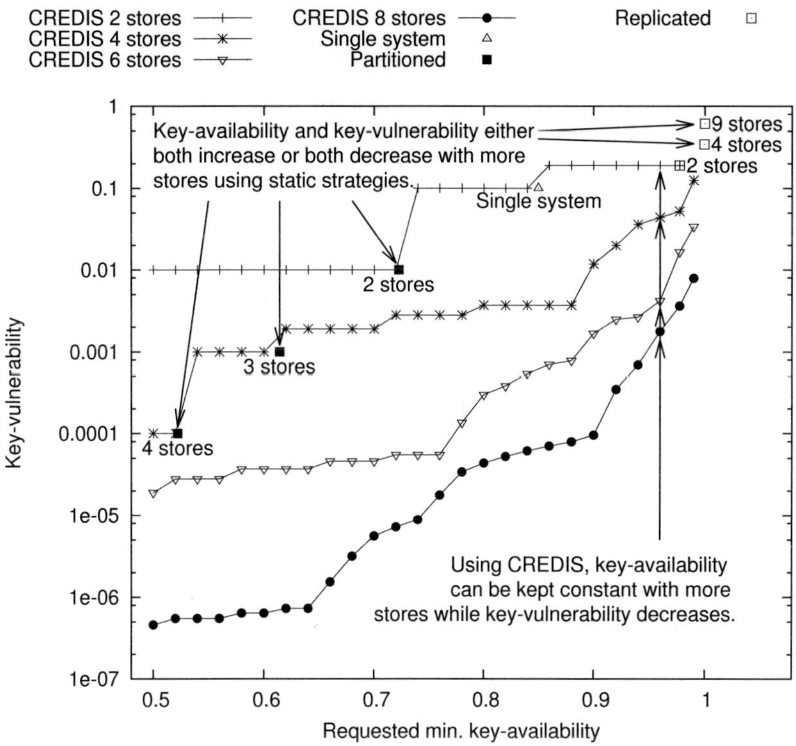

Figure 5.7: Simple strategies vs. Credis assuming store-availability 0.85 and store-vulnerability 0.1, optimization criterion: minimize key-vulnerability [KMH13].

In real scenarios, a reasonable balance between key-availability and key-vulnerability is required. In most real scenarios, a solution that is hardly available and nearly invulnerable is not deployable, just as a solution that is highly available but insecure. Credis optimally balances key-availability and key-vulnerability based on the deployment scenario requirements with regard to the maximum acceptable key-vulnerability or the minimum acceptable key-availability. Figure 5.7 shows that Credis minimizes the key-vulnerability as much as possible without violating the minimum key-availability boundary condition. Compared to the naive solution of storing the key at a single store (*single system* in Figure 5.7) Credis achieves solutions that are by order of magnitude better. If it is necessary to satisfy the specified boundary conditions, Credis can reach the same extreme solutions as the static approaches *partitioned* and *replicated*, which represent upper and lower bounds for achievable key-availability and key-vulnerability, respectively (e.g., *partitioned* with 2 stores vs. Credis with 2 stores and a minimum key-availability of 0.7225).

"We additionally measured CREDIS solutions for key-availability maximization scenarios with varying maximum key-vulnerability requirements. The results are

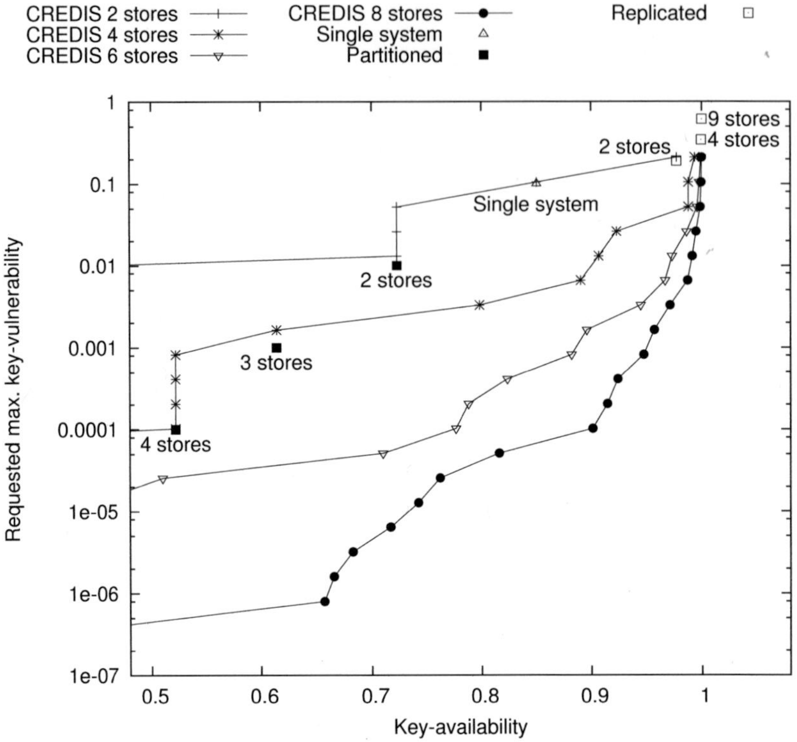

Figure 5.8: Simple strategies vs. Credis assuming store-availability 0.85 and store-vulnerability 0.1, optimization criterion: maximize key-availability [KMH13].

shown in Figure 5.8 and are very similar to those for key-vulnerability minimization scenarios with varying minimum key-availability requirements that are shown in Figure 5.7. The slight differences can be explained as follows. In Figure 5.7 the key-vulnerability of the optimal solutions is depicted in on the y-axis and the x-axis depicts the *required* minimum key-availability. Thus, the actual key-availability of each solution is greater or equal to that depicted on the x-axis. Analogously, in Figure 5.8, the key-availability of the optimal solutions is depicted on the x-axis and the *required* maximum key-vulnerability on the y-axis. The actual key-vulnerability of each solution is less or equal to that." [KMH13]

Performance of Finding Key Fragment Distributions

Parts of the section on the performance evaluation of Credis are quoted verbatim from our original publication of Credis [KMH13]. To evaluate the performance of our proposed ILP approach and the heuristics to find an optimized key fragment distribution, we compare the average execution time of the heuristics with the average execution time of the ILP approach. For each number of stores that is depicted in Figure 5.9 we generated 1000 random scenarios and measured the average solving time

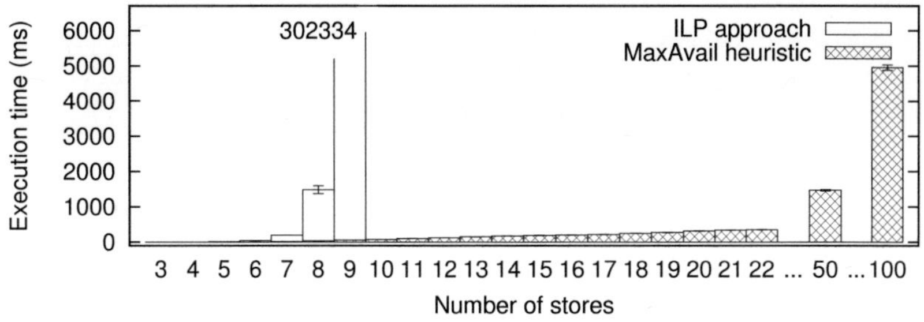

Figure 5.9: Execution time: ILP-approach vs. MaxAvail heuristic (95% confidence interval) [KMH13].

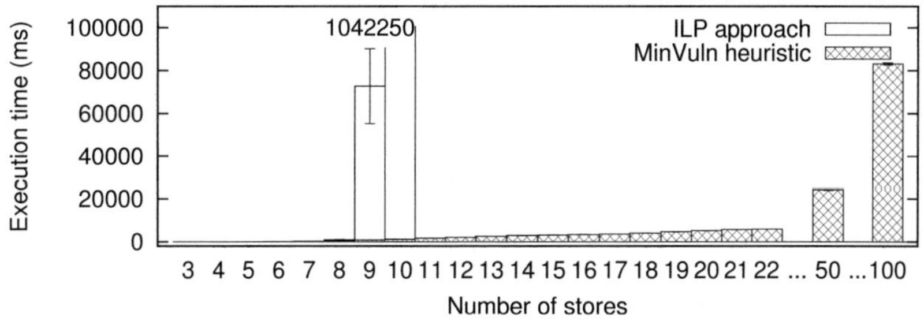

Figure 5.10: Execution time: ILP-approach vs. MinVuln heuristic (95% confidence interval) [KMH13].

of both the heuristics and the ILP approach. For each scenario, the store-availability was independently and uniformly chosen from the range $(0.7, 1]$ for each store. The store-vulnerability was independently and uniformly chosen from the range $(0, 0.2]$. In case of the MaxAvail heuristic the deployment scenario requirements contained a maximum key-vulnerability boundary condition that was uniformly chosen from the range $(0, 0.1]$. In case of the MinVuln heuristic a minimum key-availability boundary condition was uniformly chosen from the range $[0.7, 1)$ and the precision parameter was set to 2^{-15}. The maximum amount of bundles was limited to 10 in the deployment scenario requirements. The experiment was executed on a machine with 4GB RAM and a 2.93GHz dual core CPU. To solve the ILP problems, the ILP solver Gurobi[8] was utilized. Our heuristics ran single-threaded, so only one CPU core was utilized in the measurements of the heuristic's execution time.

The results of our execution time measurements for the MaxAvail heuristic are shown in Figure 5.9. Notice that solving the problem by using the ILP approach is possible in reasonable time up to around 9 stores (avg. execution time: 302 seconds, not shown in the graph). The execution time of the ILP approach increases exponentially with a rising number of stores. This is due to the size of the event relevance graph introduced in Section 5.4.3 that depends exponentially on the number of stores and

8 http://www.gurobi.com [last visited on March 2015]

the fact that solving ILP problems in general cannot be done in polynomial time. The execution time of the MaxAvail heuristic remains below 1 second even for problem instances with more than 20 stores and does not increase exponentially. Furthermore, even for large scenarios with 100 stores, the execution time of the MaxAvail heuristic remained moderate at an average of 5 seconds.

The execution time measurements of the MaxAvail heuristic are shown in Figure 5.10. While the MinVuln heuristic performs worse than the MaxAvail heuristic, large problem instances still can be solved in acceptable time. Furthermore, the ILP approach does not scale well for large key-vulnerability minimization problems either. The MinVuln heuristic can solve scenarios with 100 stores on average in 83 seconds. We assume that even bigger scenarios can be solved in acceptable time, even though as of today we cannot imagine such scenarios.

Security

The security of a distributed credential repository, i.e., whether the confidentiality of the outsourced secrets can be compromised, depends on the key-vulnerability of the key fragment distribution. Note that even if an attacker compromises all stores of a fragment, she still has to perform a successful password guessing attack to break the utilized T-PAKE protocols in order to access the secrets. Depending on the strength of the user's password this can be a difficult task for an attacker. The security of a Credis credential repository inherently depends on the security of the used security mechanisms, i.e., the T-PAKE protocols. A security analysis of these protocols is beyond the scope of this thesis.

Optimality

The **ILP approach** of Credis guarantees the optimality of the determined key fragment distribution with regard to the optimization criterion based on the following two assumptions:

- The store-availability and the store-vulnerability are correctly specified by the user.
- The store-availability and store-vulnerability of one store is independent from the store-availability and the store-vulnerability of other stores.

We methodically reduced the problem of finding an optimal key fragment distribution for given deployment scenario requirements on an ILP problem and showed how a key fragment distribution can be derived based on the ILP solution to the ILP problem. Credis makes use of ILP solvers that guarantee to determine the optimal solution for a formulated ILP problem. Thus, we claim that Credis determines optimal key fragment distributions for given deployment scenario requirements.

The version of Credis that is introduced in this chapter assumes that the store-availability and store-vulnerability of one store is independent from the store-availability and the store-vulnerability of other stores. This assumption does not hold in some scenarios. Credis can be extended to also support interdependent store-availability and store-vulnerability probabilities. The ILP model presented in Section 5.4.3

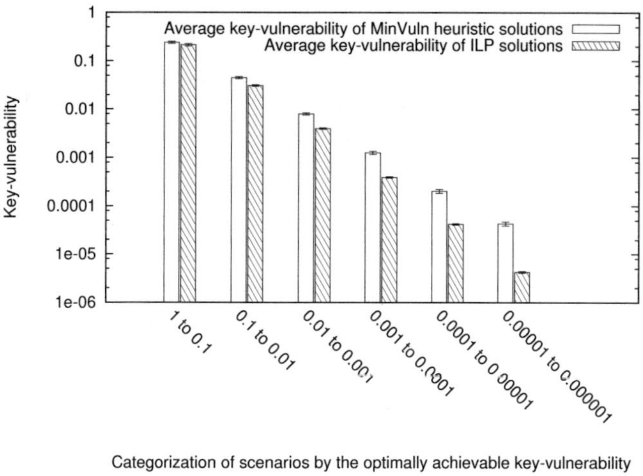

Figure 5.11: Average quality of the MinVuln heuristic solutions vs. the average quality of optimal ILP solutions (with 95% confidence interval).

can support interdependent store-availability/store-vulnerability of different stores by calculating the occurrence possibilities for each node in the event relevance graph in Figure 5.3 based on conditional probabilities.

To evaluate the **heuristics** with regard to the optimality of the produced key fragment distributions, we determined the optimal solutions of small and medium sized problem instances using the ILP algorithm and compared the solutions of the heuristics to them. We measured random scenarios with 3, 4, 5, 6, 7, and 8 stores. For each number of stores we generated and measured 10000 random scenarios. The deployment scenario requirements have been generated in the same way as for the performance evaluation.

In Figure 5.11, a comparison between the average optimal ILP solutions and the MinVuln heuristic solutions is shown for various ranges of the optimally achievable key-vulnerability in a scenario. The results show, that the MinVuln heuristic does not achieve the key-vulnerability of the ILP approach and generates slightly suboptimal solutions. This is to be expected, as the ILP solutions are optimal and no better solution exists. The quality of the MinVuln heuristic solutions nearly does not deviate from the optimal solutions for scenarios with a possible key-vulnerability in the range of 0.1 to 1.

The gap between the MinVuln heuristic and the optimal ILP solutions seems to widen for scenarios with a smaller achievable key-vulnerability. On average the generated scenarios in which a smaller key-vulnerability can be achieved contained more stores than scenarios in which only a higher key-vulnerability can be achieved. Thus, the *relative* error, i.e., the deviation of the heuristic solution to the optimal solution in relation to the achievable key-vulnerability, seems to increase if more stores are contained in the scenario. However, notice that the scale is logarithmic and the absolute error actually decreases for a smaller possible key-vulnerability. Thus, the *absolute* error that is induced by the heuristics decreases with an increasing number of stores. This is also indicated in Figure 5.12 that shows the measured dependency of the average absolute error of the heuristics on the number of stores in the generated scenarios. Even if the relative error that is induced by the heuristics can become bigger for scenarios with more stores, the absolute error can be expected to decrease for more stores. Thus, we argue that the heuristics produce solutions that are accurate enough to be applied in practice.

The results for the MaxAvail heuristic are shown in Figure 5.13 and can be interpreted analogously.

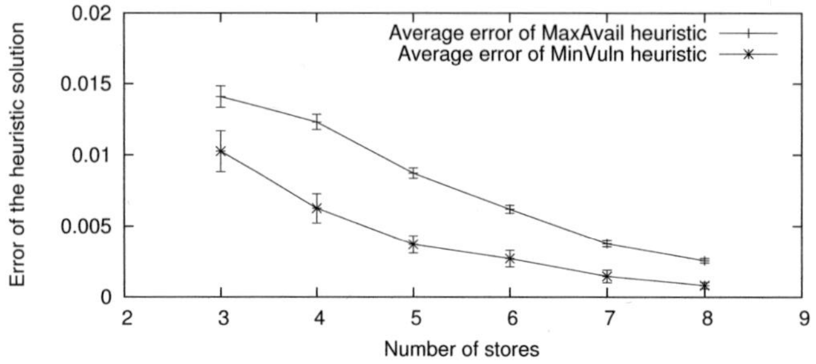

Figure 5.12: Absolute error: ILP-approach vs. greedy heuristics (with 95% confidence interval) [KMH13].

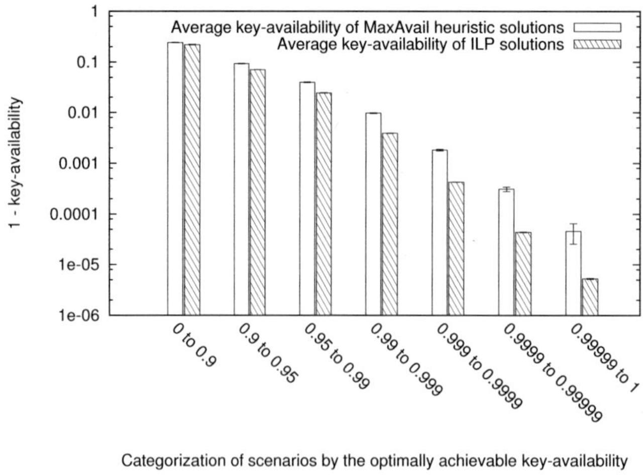

Figure 5.13: Average quality of the MaxAvail heuristic solutions vs. the average quality of optimal ILP solutions (with 95% confidence interval).

Discussion: Usability and Limitations

To derive an optimized key fragment distribution, the store-availability and store-vulnerability of each store as well as the requirements in terms of key-availability and key-vulnerability have to be specified. Developing methodologies on how to assess the store-availability and the store-vulnerability of a given store and the requirements of given scenarios is beyond the scope of this thesis. However, we provide initial ideas on how Credis' input parameters can be determined for a given scenario.

Determining requirements and store properties requires expert knowledge and is error prone. However, the determination of the requirements and store properties can be "outsourced" to experts. For instance, established institutions such as the National Institute of Standards and Technology (NIST)[9] or European Union Agency for Network and Information Security (ENISA)[10] could make recommendations on the store-availability and store-vulnerability of common store types and requirements with regard to key-vulnerability and key-availability. They already provide recommendations for other security parameters such as the length of cryptographic keys. The recommendations on the required key-vulnerability and key-availability as well as store-vulnerability and store-availability can be combined to standard scenarios that can be chosen by the user and serve as input for Credis.

Recommendations on the specification of the store-availability can be determined by measuring the percentage of time a store is available. One way to simplify the specification of the store-vulnerability is to create user guidelines based on empirical studies (e.g., "the probability of a mobile phone becoming unnoticed compromised in a timespan of 1 year amounts to 10%" [KMH13]). As a basis for such guidelines, approaches from the risk management domain such as attack trees [MLY05, Sch11] can be used to estimate store-vulnerability values. Furthermore, it is possible to specify rough estimations of store-vulnerability. For instance, one could specify store-vulnerability in a qualitative way and translate the qualitative values on quantitative ones (e.g., store-vulnerability can be either "low" = 30%, "medium" = 60% or "high" = 90%). In particular in the risk assessment domain, a prevalent way of dealing with uncertain input parameters is to specify them qualitatively and accept the subjective outcome [ISO13]. Like it is the case with risk assessment, qualitative input parameters for Credis will result in suboptimal solutions. However, in a majority of cases these solutions will be better than those that are based on no information at all on the input parameters.

"Regarding the specification of the requirements with regard to key-availability and key-vulnerability, it is possible to aid the user by presenting the range of valid values for the minimum key-availability and the maximum key-vulnerability that result in a solvable problem. The maximum reachable key-availability (and the maximum key-vulnerability) is determined by the key fragments distribution that *replicates* the key on *all* stores. The minimum key-availability (and the maximum key-vulnerability) is determined by the key fragments distribution that *fragments* the key on *all* stores. The user may choose a minimum key-availability and a maximum key-vulnerability from these ranges." [KMH13]

Once a key fragment distribution is determined, the user only has to enter her password to retrieve the secrets. This can be integrated with existing applications based on existing APIs in future work. For instance, the "Cryptography API: Next Generation

9 http://www.nist.gov/ [last visited on March 2015]
10 http://www.enisa.europa.eu/ [last visited on March 2015]

(CNG)"[11] can be used to seamlessly integrate the Credis approach with Microsoft Windows. The problem of establishing connections to the stores (including the lookup of network addresses) is independent of Credis and part of the implementation. For instance, future work can explore the possibility to lookup all stores that are part of the distributed credential repository based on distributed hash tables (DHTs).

An inherent problem of credential repositories and other approaches for the management of credentials constitutes the fact that a single password can be used to access multiple credentials. If the terminal of the user is compromised at the time when the password is entered, so are the secrets that can be accessed via this password. This especially constitutes a security issue if credentials require different protection levels (e.g., credit card PINs vs. chat messenger passwords). This security issue is inherent to the problem setting of simplifying the management of credentials. Credis allows to categorize credentials by their required protection level. For instance, it is possible to define distinct credential repositories for "high-risk" credentials and "low-risk" credentials. Thus, the user can enter his password for the "low-risk" repository at a terminal and access the according credentials without risking that "high-risk" credentials become compromised.

5.5 Conclusions

To facilitate the construction of distributed credential repositories that can be used to outsource and access credentials from arbitrary devices we proposed the Credis approach. Credis automatically generates a distributed credential repository that satisfies the individual deployment scenario requirements with regard to the availability and the vulnerability of the stored secrets. Thus, Credis allows to tune secret confidentiality for secret availability and vice versa. We approached the problem of designing a tunable approach for distributed credential repositories as follows.

- We **identified a suitable security mechanism** to build distributed credential repositories. T-PAKE protocols enforce confidentiality based on multiple potentially compromised parties by splitting up secrets among them. We showed that there exists a trade-off between the confidentiality and the availability of the secrets in a distributed credential repository.

- We **showed how the trade-off between confidentiality and availability can be tuned** in a fine grained manner by splitting up the secrets multiple times on different sets of stores. We showed that splitting up and distribute the secrets in such a way that given deployment scenario requirements are satisfied is not trivial and **determining a strategy on how to split the secrets constitutes an NP-hard problem.**

- We **formalized the optimization problem** of finding a secret splitting strategy that optimally matches the deployment scenario requirements.

[11] http://msdn.microsoft.com/en-us/library/windows/desktop/bb204778%28v=vs.85%29.aspx [last visited on March 2015]

- We **modeled the optimization problem as an ILP problem** to enable the use of ILP solvers. For bigger scenarios this approach does not scale, so we **proposed heuristics to approximate the optimal solutions** in reasonable time. We evaluated the performance and the quality of the heuristics and found them perform well for large scenarios with more than 100 stores.

We showed that Credis affects usability, as it requires users to specify the availability and vulnerability properties of the stores as well as the requirements that the credential repository has to satisfy in terms of availability and vulnerability. Thus, in deployment scenarios in which the disclosure of secrets induces low costs, the user might not apply Credis as she considers the risk of secret disclosure as disproportional to the overhead of specifying key- and store-vulnerability. However, in high-risk scenarios where assets are highly valuable, reducing the probability of secret disclosure is paramount and Credis constitutes a valuable approach. For instance, one example for a high risk scenario can be Certification Authorities that have to maintain the private key for root certificates[12]. Furthermore, Credis constitutes a foundation for the future development of more usable distributed credential repository approaches. If, for instance, recommendations are available on how to choose key- and store-vulnerability in the future, it will be interesting to apply Credis in low risk deployment scenarios as well.

The insights gained with Credis are not limited to distributed credential repositories. Every secret sharing scheme suffers from the trade-off between availability and confidentiality. Furthermore, the proposed methodology can even be applied for non-security related applications. For instance, in RAID systems there exists a trade-off between availability and performance that can be formalized similar to the trade-off between confidentiality and availability in future work.

[12] https://www.vasco.com/company/about_vasco/press_room/news_archive/2011/news_diginotar_reports_security_incident.aspx [last visited on March 2015]

6
Conclusions and Outlook

In many data outsourcing scenarios, approaches have to enforce security properties by applying security mechanisms. Yet, applying security mechanisms can negatively affect quality characteristics such as efficiency or usability. To reduce such negative effects, approaches have to be tailored to satisfy the security requirements of the given deployment scenario by enforcing only the required security properties via security mechanisms. Such specialized approaches reduce negative effects with regard to non-security quality characteristics whenever possible. We have shown that finding a suitable security mechanism combination can constitute a challenge in problem settings like database and credential outsourcing. Furthermore, deployment scenario requirements are manifold and may change over time. A single static approach that enforces specific security properties and disregards others to enhance quality characteristics like efficiency cannot satisfy the requirements of many deployment scenarios and is not suited to address requirements that change over time. Tunable approaches that allow to adapt their security properties without having to be re-designed and re-implemented can be tuned to the individual requirements of each deployment scenario and, thus, are deployable for a wide range of deployment scenarios.

In this thesis, we investigated how tunable approaches can be built that allow to trade security properties for other quality characteristics such as efficiency and usability. Tunable approaches can be used in a large variety of deployment scenarios and do not have to be replaced in case the requirements change over time. We built tunable approaches that address domain-specific problems in the domains of database outsourcing, federated identity management, and credential repositories. By doing that, we addressed the following research question: *How can security characteristics be made tunable to enable deployable data outsourcing approaches?* In the following, we summarize our contributions, draw conclusions, and give an outlook on how the results of this thesis can act as a foundation for future work.

We provided a **conceptual framework for tunable security** that puts the concept of tunability in context with other fields such as software quality, secure system development, and risk management. Furthermore, we provided a **methodology on how to build tunable approaches**. We applied this methodology to build approaches that address domain specific problems in the domains of database outsourcing, federated identity management, and credential repositories. To apply the methodology we combined methods from operations research, policy-based management, existing security mechanisms, and existing languages to express security requirements (including anonymity notions and confidentiality constraints). In particular, the application of the methodology resulted in the following major insights.

The developed approaches show that **tuning the security characteristics of an approach amounts to a constraint satisfaction problem** of finding a set of security mechanisms that satisfy the deployment scenario requirements. This constraint satisfaction problem can be perceived as an optimization problem in case specific quality properties have to be optimized. The complexity of such optimization problems strongly depends on the problem setting. We showed how operations research methodologies can be applied to solve these optimization problems when building tunable approaches and how they can be used to isolate conflicting requirements.

Furthermore, the proposed tunable approaches show that **even characteristics that appear to be binary options can be tuned in a fine-grained manner** in many cases. For instance, confidentiality can be perceived as a binary option in the sense that either the confidentiality of data is preserved or not. Our approaches show that confidentiality can an also be tuned in a fine-grained manner by 1) specifying confidentiality requirements with more detail (see confidentiality constraints of Securus, Section 3.5), 2) considering against which attackers confidentiality is enforced (see the attacker model taxonomy of CPIs, Section 3.4.3), or 3) mapping a metric to confidentiality (see key-vulnerability of Credis, Section 5.4).

Besides these generic insights, the application of the methodology resulted in contributions that are specific for the domains of secure database outsourcing, secure identity outsourcing, and secure credential outsourcing. The proposed approaches for the investigated problem settings have shown to be viable alternatives to avoiding the problem setting by not outsourcing data or accepting solutions that are sub-optimal due to the fact that they are not tailored to the deployment scenario requirements. The problem solutions and insights that are provided by this thesis for each of these domains are summarized in the following.

Secure database outsourcing: In the database outsourcing problem setting, we investigated the trade-off between confidentiality and efficiency by developing **a taxonomy for CPI approaches as well as a methodology to categorize CPIs** and using it to conduct a **wide range survey of CPI approaches** that can be used to enforce confidentiality in database indexes. To our knowledge, this constitutes the first methodology that investigates the interdependencies between CPI security properties, attacker capabilities, attacker background knowledge, and supported query functionality of

CPIs. The CPI survey shows that CPIs vary in terms of their guaranteed level of protection, their supported query functionality, and their efficiency. We provided an overview of the **tunable Securus approach** and proposed extensions to find the efficiency-optimized set of CPIs that satisfy the requirements of a given deployment scenario. Securus allows to express confidentiality requirements by specifying confidentiality constraints and is able to leverage assertions on the assumed attacker's knowledge to make use of more efficient CPIs. We formalized the optimization problem of finding an efficiency-optimized set of CPIs that enforces the specified requirements and mapped it on an ILP problem which Securus solves based on established ILP solvers. Based on the solution, Securus automatically generates a mediator software component that enforces the requirements and can be used to outsource data and subsequently query the outsourced data. If the specified requirements are not feasible, Securus isolates the set of requirements that stand in conflict to each other and presents them to the user. To our knowledge, Securus constitutes the first extensible framework that distinguishes between attackers with a varying degrees of background knowledge, can provide strict confidentiality guarantees, uses CPIs to satisfy query requirements while not requiring the user to have cryptographic expert knowledge on CPIs, leverages non-colluding storage providers for increased efficiency if available, and aids the user in detecting and resolving conflicting requirements.

Based on our CPI assessment work, it can be concluded that **CPIs differ significantly with regard to efficiency and their ability to enforce data confidentiality** against various attacker models. Thus, in order to use them to securely outsource databases and satisfy deployment scenario requirements, either CPI expert knowledge is required or the process of choosing CPIs has to be automated by an approach like Securus. Securus query performance measurements based on the TPC-C benchmark showed that **enforcing confidentiality of databases is feasible** with regard to efficiency in many database outsourcing scenarios if CPIs are applied selectively to protect only what is necessary. Furthermore, the measurements showed that other approaches that generally apply CPIs on all outsourced data induce a significant efficiency overhead which can render database outsourcing infeasible in many deployment scenarios. Furthermore, we showed that the general problem of **finding an optimal CPI combination constitutes an NP-hard optimization problem**. However, for the investigated scenarios this problem can be solved in acceptable time, for instance, by using the ILP problem formalization of Securus.

Furthermore, we contributed the **Dividat approach** to build efficiency models for anonymized databases. We exemplarily used Dividat to investigate the trade-off between ℓ-diversity requirements and efficiency. Finding a database indexing strategy that is tuned to satisfy the anonymity requirements of a deployment scenario and optimizes efficiency at the same time constitutes a challenging optimization problem that is hard to grasp. We applied Dividat to build efficiency models for anatomized ℓ-diversified databases and showed how based on these efficiency models the optimization problem of finding an efficiency-optimized indexing strategy that does not violate ℓ-diversity requirements can be formalized.

We compared the paradigms of confidential and anonymized database outsourcing and can conclude that it can be beneficial with regard to query latency and storage overhead to make use of CPIs to satisfy anonymity requirements in specific deployment scenarios. However, in other deployment scenarios it can be more beneficial to rely on anonymization techniques. For instance, CPIs can only be applied to improve the performance of evaluating specific query elements such as equality conditions while anonymization techniques can improve the performance of evaluating arbitrary query elements. Thus, ideally an approach should be able to make use of both anonymization techniques and CPIs to satisfy anonymity requirements. We conducted a first step towards this vision by providing **a concept on how the Dividat efficiency models can be used to integrate anonymity requirements in Securus.**

Secure identity outsourcing: When outsourcing identity providers to highly available cloud providers, multiple security characteristics have to be enforced if the cloud provider is assumed to be malicious. We showed that which security characteristics have to be enforced strongly depends on the trust relationships within an identity federation. We **identified possible trust models within federations** and analyzed real federations to confirm their relevance in practice. Deployment scenario requirements for federated identity management approaches in general and identity provider outsourcing approaches in particular strongly depend on the federation trust model of the deployment scenario. Thus, an approach that has to be deployable in deployment scenarios with different trust models should be tunable with regard to security properties in order to affect non-security quality characteristics as little as possible. To address the problem of securely outsourcing identity providers we proposed the **tunable Occasio approach** that relies on security mechanisms to enforce security characteristics like the confidentiality and integrity of the outsourced identity data. Occasio can be tuned to increase efficiency, usability, and availability by avoiding the use of security mechanisms whenever the deployment scenario requirements allow for it. This is possible due to the fact that the security mechanisms Occasio relies on can be applied independently from each other.

Based on Occasio, we conclude that **securely outsourcing identity providers to external parties is feasible** with regard to efficiency. However, enforcing security properties also affects other quality characteristics. We showed that **the trade-off between freshness and availability of identity information as well as a trade-off between usability and confidentiality of user passwords are inherent** to the problem setting, regardless of the used approach. Other trade-offs such as the trade-off between compatibility of the approach and freshness of the identity data originate from the boundary conditions imposed by the current environment, i.e., the existing systems that are already deployed and operational. Some of these boundary conditions can be avoided in the future, for instance, by including the support for MHT-based signatures in current federated identity management standards. Occasio shows that trade-offs can be addressed even at this point in time by tunable approaches that apply security mechanisms only if necessary.

Secure credential outsourcing: Credential repositories can be used to securely trade weak passwords for strong secrets based on the assumption that the party that hosts the credential repository is not compromised and cannot make arbitrarily many password guesses. The risk of password guessing attacks on credential repositories that are hosted by potentially compromised parties can be mitigated by using existing secret-sharing protocols. Secret sharing protocols inherently induce a trade-off between confidentiality and availability. We proposed the **tunable Credis approach** that allows to specify availability and confidentiality requirements via probabilities of the stored secrets being available or compromised. Credis optimizes availability or confidentiality without violating the specified requirement for the other quality characteristic. We formalized the according optimization problem and mapped it to an ILP problem. Furthermore, we provided heuristics to solve larger instances of the optimization problem for which solving the ILP problem results in too much computational overhead.

Credis showed that **it is possible to tune secret availability and secret confidentiality in a more fine-grained manner** than traditional threshold secret sharing schemes allow for. **Determining a secret-sharing strategy that is optimal with regard to availability and confidentiality requirements is NP-hard**. However, the problem can be solved in acceptable time for up to 9 secret-sharing parties, for instance, based on the ILP problem formalization of Credis. An important direction of future work is to simplify the specification of deployment scenario requirements.

The contributed tunable approaches show that tunable approaches can be built for many problem settings. The proposed approaches can be considered a blueprint and an inspiration for future approaches that allow to tune quality characteristics in other problem domains. In particular, we consider allowing the user to specify requirements in the form of policies and using operations research methodologies such as ILP to automatically determine security mechanisms that optimally enforce the specified requirements as generic enough to be applicable to many other problem settings. For an outlook of future work in each problem setting domain, we refer to the conclusions of the corresponding chapter. Our work acts as a foundation for further domain-independent research and leads to the following questions.

How can deployment scenario requirements be determined? The proposed tunable approaches automatically adapt their properties to satisfy specified requirements. While we provided initial ideas on how these requirements can be determined, the process of determining requirements for a given deployment scenario is subject to further research. For instance, one major problem when specifying requirements is to assure that they are not conflicting, i.e., that they can be satisfied. The approaches we proposed in this thesis constitute a first step towards solving this problem by being able to aid the user in isolating conflicting requirements.

Can the concept of "tunability" be applied on other policy layers as well? Tunable approaches like the contributed ones constitute a first step towards the vision that is advocated by the policy-based management paradigm, i.e., automatically configuring systems based on policies that are as abstract as possible. In this vision, tunable approaches address the final step of mapping policies on enforcement mechanisms.

However, we argue that the concept of tunability can be applied on the policy refinement process of mapping high-level requirements on low-level requirements as well. For instance, the requirement A="data confidentiality has to be preserved" can be refined to the requirement A'="confidentiality constraints `[X, Y]` and `[Y, Z]` have to hold". If requirement A does not exist, the requirement A' does not have to be enforced. In future work, it has to be researched if operations research methodologies can be applied to this problem in a similar way as they were applied in this thesis.

How can risk management be integrated with tunable approaches more closely? With regard to the policy refinement process, it has to be researched how risk management can be integrated more closely with tunable approaches. On an intuitive level, security mechanisms enforce security characteristics by decreasing the probability of a security breach. A risk manager has to decide for each deployment scenario which probability is sufficient. For instance, to enforce the requirement "data confidentiality has to be preserved", the risk manager has to decide which requirements a system has to satisfy in order to reach a security breach probability that is low enough. At the same time, the risk manager has to be aware of the costs that come with the satisfaction of these requirements with regard to other quality characteristics. To address this problem, our approaches can be combined with risk management methods such as attack trees in future work. For instance, by combining Securus with attack trees it can be determined by how much efficiency increases if an increased level of risk is accepted. Thus, not only trade-offs between security and non-security quality characteristics can be explored but also trade-offs between risk and non-security quality characteristics.

A

Appendix: Securus

A.1 Proof: NP-Hardness of Satisfying Policy Profiles

Theorem 1. *Finding a feasible attribute allocation that satisfies a policy profile is NP-hard.*

Proof. We reduce the NP-complete problem of 3-coloring an undirected graph on the problem of finding an attribute allocation that enforces a policy profile.

The Graph 3-Coloring Problem:

> **Given:** An undirected graph $G = (V, E)$ with a set of nodes V and a set of edges E
>
> **Wanted:** A coloring of the nodes using at most three colors, so that two nodes that are connected via an edge do not have the same color.

Reduction: For a given problem instance of the 3 color problem, define a policy profile as follows:

1. Define the set of available CSPs $\mathcal{S}$ so that $|\mathcal{S}| = 3$ holds.
2. Generate a unique attribute a_i for each node $v_i \in V$ and add it to the attribute set in the policy profile.
3. Add a QP "SELECT ... WHERE a_i=?" for each attribute a_i to the policy profile.
4. For each edge $(v_x, v_y) \in E$, add a CC $[a_x, a_y]$ to the policy profile where a_x and a_y are exactly those attributes that where generated for v_x and v_y in step 2.

Find an attribute allocation that satisfies the policy profile. This attribute allocation can be transformed to a solution of the 3-coloring problem as follows:

1. Map each color to one of the CSPs in $\mathcal{S}$.
2. For each node v_i choose one CSP that may store attribute a_i according to the attribute allocation. Color the node v_i with the color that is mapped to the chosen CSP.

Note that both the reduction of a 3-coloring problem instance on a policy profile and the transformation of the attribute allocation on the 3-coloring solution are doable in polynomial time.

Proof of correctness (policy profile feasible ⇒ 3-coloring problem feasible): If a policy profile is feasible, solving the ILP problem results in a feasible attribute allocation. Lets assume that our reduction produces a feasible attribute allocation for a policy profile but an incorrect 3-coloring solution. A 3-coloring solution can be incorrect because of two reasons: Either two nodes that are connected via an edge have the same color or at least one node is not colored at all.

Case 1 - Two nodes v_x and v_y that are connected via an edge (v_x, v_y) have the same color: The edge (v_x, v_y) implies that there exists a CC $[v_x, v_y]$ in the policy profile. The fact that both nodes v_x and v_y have the same color implies that the attributes a_x and a_y that v_x and v_y were mapped on are both stored by a single CSP. This in turn would violate the CC $[a_x, a_y]$. Thus, the attribute allocation does not constitute a feasible solution for the generated policy profile. This constitutes a contradiction.

Case 2 - A node v_x is not colored at all: The node v_x can only be not colored if no CSP exists in $\mathcal{S}$ that stores attribute a_x. Thus, the attribute allocation is no feasible solution for the generated policy profile, as the generated QP "SELECT ... WHERE a_i=?" requires that a_x is at least stored by one CSP.

Proof of completeness (3-coloring problem feasible ⇒ policy profile feasible): Lets assume that a solution exists for a given 3-coloring problem instance. An attribute allocation that satisfies the generated policy profile can be constructed as follows:

1. Map each color to one of the CSPs in $\mathcal{S}$.
2. Store each attribute a_i at the CSP that is mapped on the color of v_i.

There are two possible reasons why an attribute allocation does not satisfy a policy profile: Either a QP is violated or a CC is violated.

Case 1 - A QP is violated by the attribute allocation: As in a feasible 3-coloring solution each node has a color, each attribute is mapped to one CSP. The generated policy profile only contains singleton QPs for each attribute. As each attribute is stored by at least one CSP, no QP is violated and this case cannot occur.

Case 2 - A CC is violated by the attribute allocation: Let us assume that a CC $[a_x, a_y]$ that is violated by the attribute allocation exists. This implies that attribute a_x and a_y

are stored by the same CSP. Thus, the corresponding nodes v_x and v_y have the same color. The existence of the CC $[a_x, a_y]$ implies that there exists an edge (v_x, v_y) in the 3-coloring problem. As the nodes v_x and v_y have the same color, this contradicts with the assertion that the 3-coloring solution is sound. □

Theorem 2. *Finding an optimal attribute allocation that satisfies a policy profile is NP-hard even if just one CSP can be used to store the outsourced data.*

Proof. We reduce the NP-complete problem of finding a minimum vertex cover of an undirected graph on the problem of finding an optimal attribute allocation that enforces a policy profile.

The Minimum Vertex Cover Problem:

> **Given:** An undirected graph $G = (V, E)$ with a set of nodes V and a set of edges E.
>
> **Wanted:** A minimum vertex cover, i.e., a subset of nodes $U \in V$ so that for all edges $(v_x, v_y) \in E$ it holds that either v_x or v_y are contained in U. There must not exist another vertex cover $U' \subset V$ so that $|U'| < |U|$ holds.

Reduction: For a given problem instance of a vertex cover problem, define a policy profile as follows:

1. Define the set of available CSPs $\mathcal{S}$ so that $|\mathcal{S}| = 1$ holds.
2. Generate a unique attribute a_i for each node v_i in V and add it to the attribute set in the policy profile.
3. Add a QP "SELECT ... WHERE a_i=?" for each attribute a_i to the policy profile.
4. For each edge (v_x, v_y), add a CC $[a_x, a_y]$ to the policy profile where a_x and a_y are exactly those attributes that where generated for v_x and v_y in step 2.
5. Add a DIC $[a_i]$ for each attribute a_i to the policy profile.
6. For each QP $q \in QP$, set the cost $c^d_{q,i}$ of evaluating the according queries based on distinguishable ciphertexts of a_i to 1. Set all other costs $c^d_{q,i}$ and $t^d_{q,i}$ to 0.

Find an attribute allocation that satisfies optimally the policy profile. This attribute allocation can be transformed to a solution of the vertex cover problem as follows:

1. For each attribute a_i that is stored as distinguishable ciphertext at the CSP, add v_i to the vertex cover set U.

Note that both the reduction of a vertex cover problem instance on a policy profile and the transformation of the attribute allocation on the vertex cover set are doable in polynomial time.

Proof (attribute allocation satisfies policy profile ⇔ U is a valid vertex cover): Storing plaintext values of a_x and a_y violates the CC $[a_x, a_y]$. Due to the specified DICs, distinguishable ciphertexts may be securely outsourced for any attribute. Protecting attributes by outsourcing order-preserving ciphertexts is not an option as no OICs are specified. Thus, to satisfy the specified QPs, either distinguishable ciphertexts or plaintext values have to be outsourced for each attribute.

A policy profile that was constructed as shown above is satisfied if for each specified CC $[a_x, a_y]$ only distinguishable ciphertexts are stored for a_x or a_y or both. This is exactly the semantic of a vertex cover. For each edge $(v_x, v_y) \in E$, v_x or v_y or both nodes have to be contained in the vertex cover. As attributes a_i of the policy profile directly map to vertices v_i of the graph and CCs $[a_x, a_y]$ directly map to edges (v_x, v_y) in the graph, the node set U constitutes a valid vertex cover exactly when the attribute allocation satisfies the policy profile.

Proof (attribute allocation is optimal ⇒ U is an optimal vertex cover): We prove the implication by contradiction. Lets assume that based on an optimal attribute allocation that protects the attributes $a_{k_1}, \ldots, a_{k_{l+1}}$, the reduction produces a non-minimal vertex cover $U = v_{k_1}, \ldots, v_{k_{l+1}}$. Thus, there exists another vertex cover $v_{u_1}, \ldots, v_{u_l} = U'$ with $|U'| = l < l + 1 = |U|$. As the vertex cover is valid, an attribute allocation that only stores distinguishable ciphertexts of the attributes $a_{u_1}, \ldots, a_{u_l}$ also satisfies the policy profile. The attribute allocation induces the cost l as each attribute for which distinguishable ciphertexts are outsourced induces the cost 1. This is a contradiction to the assumption that the original attribute allocation that stores distinguishable ciphertexts of $l + 1$ attributes $a_{k_1}, \ldots, a_{k_{l+1}}$ is optimal as this attribute allocation induces the costs $l + 1$.

Proof (U is an optimal vertex cover ⇒ attribute allocation is optimal): We prove the implication by contradiction. Let us assume that an optimal vertex cover $U = v_{k_1}, \ldots, v_{k_{l+1}}$ of size $l + 1$ has been derived from a non-optimal attribute allocation that protects the attributes $a_{k_1}, \ldots, a_{k_{l+1}}$. Thus, there exists another attribute allocation that only protects l attributes $a_{u_1}, \ldots, a_{u_l}$ and that still satisfies the defined policy profile. As the attribute allocation satisfies the policy profile, $U' = v_{u_1}, \ldots, v_{u_l}$ constitutes a valid vertex cover. Thus, a vertex cover of the size l exists which contradicts with the assumption. □

A.2 Policy Transformation: Securus' ILP Problem

The Securus ILP problem formalizes the conditions that a feasible attribute allocation has to comply with to satisfy a policy profile. In the following we present the formalization of Securus' ILP problem. To present the ILP formalization we first introduce constants which represent the specified policy profile and define functions that allow us to present the ILP problem formalization in a compressed way. We introduce the meaning of the free variables of the ILP problem and list the formalization of the ILP target function as well as the ILP constraints. After that, we explain the ILP formalization by showing how the conditions that we presented in Section 3.5.3 map to the ILP problem constraints.

Constants:

A : Set of attributes.
CC : Set of confidentiality constraints (CCs).
OIC : Set of order-preserving inference constraints (OICs).
DIC : Set of distinguishable inference constraints (DICs).
QP : Set of access policies (QPs).
S : Set of storage providers (CSPs).
$c^o_{q,i}$: Cost of evaluating QP q based on order-preserving ciphertexts of attribute i.
$c^d_{q,i}$: Cost of evaluating QP q based on deterministic ciphertexts of attribute i.
$c^i_{q,i}$: Cost of evaluating QP q based on indistinguishable ciphertexts of attribute i.

Functions:

$$o(i,j) = \begin{cases} O_{i,j} \text{ if } i \notin OIC \\ 0 \text{ else} \end{cases}$$

$$d(i,j) = \begin{cases} D_{i,j} \text{ if } i \notin DIC \cup OIC \\ 0 \text{ else} \end{cases}$$

$s(q)$: Set of selected attributes in QP q.
$e(q)$: Set of attributes on which equality selection conditions are evaluated in QP q (Type 1,4 query component).
$r(q)$: Set of attributes on which range selection conditions are evaluated in QP q (Type 2 query component).
$l(q)$: Set of attributes on which like selection conditions and other conditions that require plaintext values are evaluated in QP q (Type 3,6,10,11 query components).
$a(q)$: Set of attributes contained in simple aggregation components in QP q (Type 5 query component).
$g(q)$: Set of GROUP BY attributes in QP q (Type 8,7 query component).
$p(q)$: Set of SORTED BY attributes in QP q (Type 9 query component).

Free variables: All variables that are presented in the following are binary variables, i.e., they can only take values 1 and 0. In the following it is explained what the value 1 means for each free variable.

$P_{i,j}$: Values of attribute i are stored at CSP j in plaintext.
$O_{i,j}$: Values of attribute i are stored at CSP j as order-preserving ciphertexts.
$D_{i,j}$: Values of attribute i are stored at CSP j as distinguishable ciphertexts.
$I_{i,j}$: Values of attribute i are stored at CSP j as indistinguishable ciphertexts.
$Z_{i,j}$: Values of attribute i are revealable by CSP j.
$Q_{q,j}$: CSP j can evaluate queries of QP q.
$o_{i,q}$: QP q is evaluated based on order-preserving ciphertexts of attribute i.
$d_{i,q}$: QP q is evaluated based on distinguishable ciphertexts of attribute i.
$i_{i,q}$: QP q is evaluated based on indistinguishable ciphertexts of attribute i.

ILP optimization problem:

$$\min w_l \cdot \sum_{q\in QP} \sum_{i\in q} (c^o_{q,i} \cdot o_{i,q} + c^d_{q,i} \cdot d_{i,q} + c^i_{q,i} \cdot i_{i,q}) + w_t \cdot \sum_{q\in QP} \sum_{i\in q} (t^o_{q,i} \cdot o_{i,q} + t^d_{q,i} \cdot d_{i,q} + t^i_{q,i} \cdot i_{i,q})$$

s.t.

$$\forall q \in QP: \quad \sum_{j\in S} Q_{q,j} \geq 1 \tag{A.1}$$
$$\forall j \in S, \forall q \in QP, \forall i \in s(q): \quad P_{i,j} + O_{i,j} + D_{i,j} + I_{i,j} - Q_{q,j} \geq 0 \tag{A.2}$$
$$\forall j \in S, \forall q \in QP, \forall i \in e(q): \quad P_{i,j} + O_{i,j} + D_{i,j} - Q_{q,j} \geq 0 \tag{A.3}$$
$$\forall j \in S, \forall q \in QP, \forall i \in r(q): \quad P_{i,j} + O_{i,j} - Q_{q,j} \geq 0 \tag{A.4}$$
$$\forall j \in S, \forall q \in QP, \forall i \in l(q): \quad P_{i,j} - Q_{q,j} \geq 0 \tag{A.5}$$
$$\forall j \in S, \forall q \in QP, \forall i \in a(q): \quad P_{i,j} + O_{i,j} + D_{i,j} + I_{i,j} - Q_{q,j} \geq 0 \tag{A.6}$$
$$\forall j \in S, \forall q \in QP, \forall i \in g(q): \quad P_{i,j} + O_{i,j} + D_{i,j} - Q_{q,j} \geq 0 \tag{A.7}$$
$$\forall j \in S, \forall q \in QP, \forall i \in p(q): \quad P_{i,j} + O_{i,j} - Q_{q,j} \geq 0 \tag{A.8}$$
$$\forall c \in CC, \forall j \in S: \quad \sum_{i\in c} Z_{i,j} < |c| \tag{A.9}$$
$$\forall j \in S, \forall i \in A: \quad P_{i,j} + o(i,j) + d(i,j) - Z_{i,j} \leq 0 \tag{A.10}$$
$$\forall q \in QP, \forall j \in S: \quad Q_{q,j} \in \{0,1\} \tag{A.11}$$
$$\forall i \in A, \forall j \in S: \quad P_{i,j} \in \{0,1\} \tag{A.12}$$
$$\forall i \in A, \forall j \in S: \quad O_{i,j} \in \{0,1\} \tag{A.13}$$
$$\forall i \in A, \forall j \in S: \quad D_{i,j} \in \{0,1\} \tag{A.14}$$
$$\forall i \in A, \forall j \in S: \quad I_{i,j} \in \{0,1\} \tag{A.15}$$
$$\forall i \in A, \forall j \in S: \quad Z_{i,j} \in \{0,1\} \tag{A.16}$$
$$\forall q \in QP, \forall i \in q: \quad o_{i,q} \in \{0,1\} \tag{A.17}$$
$$\forall q \in QP, \forall i \in q: \quad d_{i,q} \in \{0,1\} \tag{A.18}$$
$$\forall q \in QP, \forall i \in q: \quad i_{i,q} \in \{0,1\} \tag{A.19}$$
$$\forall q \in QP, \forall i \in q, \forall j \in S: \quad o_{i,q} - Q_{q,j} - O_{i,j} \geq -1 \tag{A.20}$$
$$\forall q \in QP, \forall i \in q, \forall j \in S: \quad d_{i,q} - Q_{q,j} - D_{i,j} \geq -1 \tag{A.21}$$
$$\forall q \in QP, \forall i \in q, \forall j \in S: \quad i_{i,q} - Q_{q,j} - I_{i,j} \geq -1 \tag{A.22}$$

The objective of the ILP problem is to find an attribute allocation, i.e., a value combination for $P_{i,j}$, $O_{i,j}$, $D_{i,j}$, and $I_{i,j}$, that optimizes the target function and satisfies the policy profile. The target function was already explained in detail in Section 3.5.4. We now summarize the conditions that have to be met to satisfy a policy profile (see Section 3.5.3) and show how they map to the ILP problem constraints.

Condition 1: For each QP at least one CSP has to store the relevant attributes in a representation that allows the CSP to execute efficiently the queries which are modeled by the QP. This condition is expressed by the ILP constraint shown in Equation A.1. For a CSP j to be able to participate in the evaluation of a query q ($Q_{q,j} = 1$), it has to be allowed to store specific attribute representations depending on the query components of q. An overview of attribute representations that are required to evaluate different query components is shown in Table 3.5. The ✓ symbol indicates that there exists a CPI that implies the according attribute representation and that allows the CSP to evaluate the according query component. To allow for a participation of a CSP j in the process of evaluating query q, the CSP has to be allowed to store attribute representation so that a CPI exists for each query component of q. These conditions are modeled in the ILP constraints that are shown in Equation A.2, A.3, A.4, A.5, A.6, A.7, and A.8. For instance, if a query q contains an equality selection condition on attribute i (type 1 query component type in Table 3.5), the CSP j has to be allowed to store plaintext values ($P_{i,j} = 1$), order-preserving ciphertexts ($O_{i,j} = 1$), or distinguishable ciphertexts ($D_{i,j} = 1$) to be able to evaluate the query q ($Q_{q,j} = 1$). This is expressed by Equation A.3. If $Q_{q,j} = 1$ holds then either $P_{i,j} = 1$, $O_{i,j} = 1$, or $D_{i,j} = 1$ has to hold for Equation A.3 to be true.

Condition 2: For all CCs it has to hold that no CSP is able to view values for all attributes that are contained in the CC. This condition is expressed by the ILP constraint shown in Equation A.9. The CC is satisfied if each CSP j is not able to reveal the values of at least one attribute i that is contained in the CC ($Z_{i,j} = 0$). The following conditions have to hold so that a CSP j cannot reveal values of an attribute i.

- **Condition 2.1:** Plaintext values of attribute i must not be stored by CSP j.
- **Condition 2.2:** If no DIC that contains the attribute i is specified, distinguishable ciphertexts of attribute i must not be stored by CSP j.
- **Condition 2.3:** If no OIC that contains the attribute i is specified, order-preserving ciphertexts of attribute i must not be stored by CSP j.

These conditions are expressed by the ILP constraint that is shown in Equation A.10. For instance, if no DIC or OIC is specified for attribute i ($o(i,j) = O_{i,j}$ and $d(i,j) = D_{i,j}$), for $Z_{i,j} = 0$ to hold it has to hold that no plaintext values ($P_{i,j} = 0$), no distinguishable ciphertexts ($D_{i,j} = 0$), and no order-preserving ciphertexts ($O_{i,j} = 0$) may be stored by CSP j.

All variables have to be binary, i.e., they may only take the values 0 or 1. This is expressed by the ILP constraints shown in Equation A.11-A.19.

Finally, the target function is based on the helping variables $o_{i,q}$, $d_{i,q}$, and $i_{i,q}$ which assert that queries of the query policy q are executed based on order-preserving, distinguishable, and indistinguishable ciphertexts of attribute i respectively. For instance, a query q has to be executed based on order preserving ciphertexts of attribute i ($o_{i,q} = 1$) if the CSP j that can evaluate the query ($Q_{q,j} = 1$) may only store order-preserving ciphertexts of attribute i ($O_{i,j} = 1$). This condition is expressed by the ILP constraint shown in Equation A.20 for order-preserving ciphertexts. Equation A.21 and Equation A.22 contain analogous conditions for distinguishable and indistinguishable ciphertexts.

A.3 Policy Transformation: Performance of SMT solvers

Up to now, we modeled the problem of finding an efficiency-optimal CPI combination as an ILP problem. The problem can also be modeled by using other representations of optimization problems. One such representation are Satisfiability Modulo Theories (SMT) [DMB11] which allow to model decision problems via logical formulas. SMT problem instances can be considered as a generalization of boolean SAT problems as they allow to replace binary variables with predicates. Furthermore, existing solvers for SMT problems like Z3[1] were extended to also support optimization objectives [BP14], which makes them suitable to be applied in the Securus problem setting.

We modeled the problem of finding an optimal CPI combination for a policy profile as an SMT problem and measured the time that is required to find an optimal solution for policy profiles of different size. We generated those policy profiles as described in Section 3.5.7. We used the SMT solver Z3 for our measurements. A comparison of the measurement results in Table A.2 with the performance measurements of the ILP approach in Table 3.6 shows that the utilized ILP solvers solve the problems more efficiently. This is due to the fact that the SMT solvers rely on solving techniques that are also used by ILP solvers (including the simplex algorithm). However, specialized ILP solvers like Gurobi are far more optimized in the sense that they make optimized decisions on which solving technique should be applied to solve a given optimization problem as efficiently as possible.

SMT solvers are able to identify *unsatisfiable cores* within an SMT problem instance. Like *irreducible inconsistent sets* in ILP problems, unsatisfiable cores can be used to isolate conflicting policies in the policy profile (see Section 3.5.5). We measured the time that is needed to isolate conflicting policies in policy profiles based on the Z3 SMT solver. The results are shown in Table A.2. The measurement results show that SMT solvers are able to identify conflicting policies more efficiently than ILP solvers. This can be explained by the fact that finding conflicting policies is closer to a decision problem (for which SMT solvers are specialized) than an optimization problem (for which ILP solvers are specialized).

Thus, in terms of efficiency it makes sense to leverage the synergies of ILP and SMT solvers by applying ILP solvers to solve Securus policy profiles and SMT solvers to isolate policy conflicts in unsolveable policy profiles.

[1] https://github.com/Z3Prover/z3

Number of policies/elements					Duration (ms)	
Attr.	QPs	CCs	ICs	SPs	mean	max
10	10	5	1	2	360.64	535.0
10	20	15	2	2	1119.90	1685.0
40	20	10	2	2	15345.70	15925.0
60	40	40	3	2	16842.17	46343.0
80	80	60	6	2	103348.32	140927.0
100	80	80	6	2	150671.96	155304.0

Table A.1: Time required to generate a mediator from policy profiles via the Z3 SMT solver.

Number of policies/elements					Duration (ms)	
Attr.	QPs	CCs	ICs	SPs	mean	max
10	10	5	1	2	24.55	495.0
10	10	5	1	3	16.86	25.0
10	20	15	2	2	17.97	54.0
10	20	15	2	3	19.93	26.0
40	20	10	2	2	32.65	83.0
40	20	10	2	3	42.03	64.0
50	40	40	3	2	27.68	36.0
50	40	40	3	3	34.83	43.0
60	40	40	3	2	28.59	35.0
60	40	40	3	3	38.36	85.0
80	80	60	6	2	38.49	55.0
80	80	60	6	3	57.58	105.0
100	80	80	6	2	44.83	63.0
100	80	80	6	3	60.86	81.0

Table A.2: Time required to find policy conflicts in unsatisfiable policy profiles of various sizes via the Z3 SMT solver.

A.4 Policy Profile for the TPCC Benchmark

To illustrate how Securus-Latin can be applied to express deployment scenario requirements, we provide a Securus policy profile for the popular TPC-C benchmark data and workload [Tra10]. "The Company portrayed by the benchmark is a wholesale supplier with a number of geographically distributed sales districts and associated warehouses. As the Company's business expands, new warehouses and associated sales districts are created. Each regional warehouse covers 10 districts. Each district serves 3,000 customers. All warehouses maintain stocks for the 100,000 items sold by the Company. [...] Customers call the Company to place a new order or request the status of an existing order. Orders are composed of an average of 10 order lines (i.e., line items). One percent of all order lines are for items not in-stock at the regional warehouse and must be supplied by another warehouse." [Tra10]

To specify a policy profile, the data model, the queries that will be executed on the data, and the confidentiality requirements have to be specified. The data model is already specified by the TPC-C specification and includes the data tables and the attributes that are contained in each table [Tra10]. Furthermore, the queries that are executed on the data are also specified by the TPC-C specification and can be copied to the policy profile. In the following, we list the policy profile we developed for the TPC-C workload and show how confidential data can be protected by specifying ICs and CCs.

```
Policy Profile tpcc{
  Tables{
    WAREHOUSE {
      W_ID : INTEGER,
      W_NAME : VARCHAR(10) ,
      W_STREET_1 : VARCHAR(20),
      W_STREET_2 : VARCHAR(20),
      W_CITY : VARCHAR(20),
      W_STATE : CHAR(2),
      W_ZIP : CHAR(2),
      W_TAX : FLOAT(4),
      W_YTD : FLOAT(12),
    }
    DISTRICT {
      D_ID : INTEGER,
      D_W_ID : INTEGER,
      D_NAME : VARCHAR(10),
      D_STREET_1 : VARCHAR(20),
      D_STREET_2 : VARCHAR(20),
      D_CITY : VARCHAR(20),
      D_STATE : CHAR(2),
      D_ZIP : CHAR(9),
      D_TAX : FLOAT(4),
      D_YTD : FLOAT(12),
      D_NEXT_O_ID : INTEGER,
    }
    CUSTOMER {
      C_ID : INTEGER,
      C_D_ID : INTEGER,
      C_W_ID : INTEGER,
      C_FIRST : VARCHAR(16),
      C_MIDDLE : CHAR(2),
      C_LAST : VARCHAR(16),
      C_STREET_1 : VARCHAR(20),
      C_STREET_2 : VARCHAR(20),
      C_CITY : VARCHAR(20),
      C_STATE : CHAR(2),
      C_ZIP : CHAR(9),
      C_PHONE : CHAR(16),
      C_SINCE : DATE,
      C_CRED : CHAR(2),
      C_CRED_LIM : FLOAT(12),
      C_DISCOUNT : FLOAT(4),
      C_BALANCE : FLOAT(12),
      C_YTD_PAYMENT : FLOAT(12),
      C_PAYMENT_CNT : FLOAT(4),
      C_DELIVERY_CNT : FLOAT(4),
      C_DATA : VARCHAR(500),
    }
    HISTORY {
      H_C_ID : INTEGER,
      H_C_D_ID : INTEGER,
      H_C_W_ID : INTEGER,
```

```
    H_D_ID : INTEGER,
    H_W_ID : INTEGER,
    H_DATE : DATE,
    H_AMOUNT : FLOAT(6),
    H_DATA   : VARCHAR(24),
  }
  NEWORDER {
    NO_O_ID : INTEGER,
    NO_D_ID : INTEGER,
    NO_W_ID : INTEGER,
  }
  ORDERS {
    O_ID : INTEGER,
    O_D_ID : INTEGER,
    O_W_ID : INTEGER,
    O_C_ID : INTEGER,
    O_ENTRY_D : DATE,
    O_CARRIER_ID : INTEGER,
    O_OL_CNT : FLOAT(2),
    O_ALL_LOCAL : FLOAT(1),
  }
  ORDERLINE {
    OL_O_ID : INTEGER,
    OL_D_ID : INTEGER,
    OL_W_ID : INTEGER,
    OL_NUMBER : INTEGER,
    OL_I_ID : INTEGER,
    OL_SUPPLY_W_ID : INTEGER,
    OL_DELIVERY_D : DATE,
    OL_QUANTITY : FLOAT(2),
    OL_AMOUNT : FLOAT(6),
    OL_DIST_INFO : CHAR(24),
  }
  ITEM {
    I_ID : INTEGER,
    I_IM_ID : INTEGER,
    I_NAME : VARCHAR(24),
    I_PRICE : FLOAT(5),
    I_DATA : VARCHAR(50),
  }
  STOCK {
    S_I_ID : INTEGER,
    S_W_ID : INTEGER,
    S_QUANTITY : FLOAT(4),
    S_DIST_01 : CHAR(24),
    S_DIST_02 : CHAR(24),
    S_DIST_03 : CHAR(24),
    S_DIST_04 : CHAR(24),
    S_DIST_05 : CHAR(24),
    S_DIST_06 : CHAR(24),
    S_DIST_07 : CHAR(24),
    S_DIST_08 : CHAR(24),
    S_DIST_09 : CHAR(24),
```

```
      S_DIST_10 : CHAR(24),
      S_REMOTE_CNT : FLOAT(8),
      S_ORDER_CNT : FLOAT(4),
      S_REMOTE_CNT : FLOAT(4),
      S_DATA : VARCHAR(50),
    }
  }

QueryPolicies{
  //The New-Order Transaction
  /////////////////////////////

  SELECT W_TAX FROM WAREHOUSE WHERE W_ID = ?;
  SELECT D_TAX,D_NEXT_O_ID FROM DISTRICT WHERE D_W_ID = ?
     AND D_ID = ?;
  SELECT C_DISCOUNT,C_LAST,C_CRED FROM CUSTOMER WHERE C_W_ID
      = ? AND C_D_ID = ? AND C_ID = ?;
  SELECT I_PRICE,I_NAME,I_DATA FROM ITEM WHERE I_ID = ?;
  SELECT S_QUANTITY,S_DIST_01,S_DIST_02,S_DIST_03,S_DIST_04,
     S_DIST_05,S_DIST_06,S_DIST_07,S_DIST_08,S_DIST_09,
     S_DIST_10,S_DATA,S_YTD,S_ORDER_CNT,S_REMOTE_CNT FROM
     STOCK WHERE S_I_ID = ? AND S_W_ID = ?;

  //The Payment Transaction
  /////////////////////////////

  SELECT W_NAME,W_STREET_1,W_STREET_2,W_CITY,W_STATE,W_ZIP,
     W_YTD FROM WAREHOUSE WHERE W_ID = ?;
  SELECT D_NAME,D_STREET_1,D_STREET_2,D_CITY,D_STATE,D_ZIP,
     D_YTD FROM DISTRICT WHERE D_W_ID = ? AND D_ID = ?;
  SELECT C_DATA,C_FIRST,C_MIDDLE,C_LAST,C_STREET_1,
     C_STREET_2,C_CITY,C_STATE,C_ZIP,C_PHONE,C_SINCE,C_CRED,
     C_CRED_LIM,C_DISCOUNT,C_BALANCE,C_YTD_PAYMENT,
     C_PAYMENT_CNT,C_DATA,C_ID,C_D_ID,C_W_ID FROM CUSTOMER
     WHERE C_W_ID = ? AND C_D_ID = ? AND C_ID = ?;
  SELECT C_DATA,C_ID,C_FIRST,C_MIDDLE,C_STREET_1,C_STREET_2,
     C_CITY,C_STATE,C_ZIP,C_PHONE,C_SINCE,C_CRED,C_CRED_LIM,
     C_DISCOUNT,C_BALANCE,C_YTD_PAYMENT,C_PAYMENT_CNT,C_DATA
     ,C_ID,C_D_ID,C_W_ID FROM CUSTOMER WHERE C_W_ID = ? AND
     C_D_ID = ? AND C_LAST = ? ORDER BY C_FIRST ASC;

  //The Order-Status Transaction
  /////////////////////////////

  SELECT C_BALANCE,C_FIRST,C_MIDDLE,C_LAST FROM CUSTOMER
     WHERE C_W_ID = ? AND C_D_ID = ? AND C_ID = ?;
  SELECT C_BALANCE,C_FIRST,C_MIDDLE,C_LAST FROM CUSTOMER
     WHERE C_W_ID = ? AND C_D_ID = ? AND C_ID = ? ORDER BY
     C_FIRST ASC;
  SELECT O_ID,O_ENTRY_D,O_CARRIER_ID FROM ORDERS WHERE
     O_W_ID = ? AND O_D_ID = ? AND O_C_ID = ? AND O_ID = (
     SELECT MAX(O_ID) FROM ORDERS WHERE O_W_ID = ? AND
     O_D_ID = ? AND O_C_ID = ?);
```

```
    SELECT OL_I_ID,OL_SUPPLY_W_ID,OL_QUANTITY,OL_AMOUNT, OL_DELIVERY_D FROM ORDERLINE WHERE OL_W_ID = ? AND OL_D_ID = ? AND OL_O_ID = ?;

    //The Delivery Transaction
    ////////////////////////////

    SELECT NO_O_ID FROM NEWORDER WHERE NO_W_ID = ? AND NO_D_ID = ? AND NO_O_ID = (SELECT MIN(NO_O_ID) FROM NEWORDER WHERE NO_W_ID = ? AND NO_D_ID = ?);
    SELECT O_C_ID FROM ORDERS WHERE O_W_ID = ? AND O_D_ID = ? AND O_ID = ?;
    SELECT SUM(OL_AMOUNT) FROM ORDERLINE WHERE OL_W_ID = ? AND OL_D_ID = ? AND OL_O_ID = ?;
    SELECT C_BALANCE,C_DELIVERY_CNT FROM CUSTOMER WHERE C_W_ID = ? AND C_D_ID = ? AND C_ID = ?;

    //The Stock-Level Transaction
    ////////////////////////////

    SELECT D_NEXT_O_ID FROM DISTRICT WHERE D_W_ID = ? AND D_ID = ?;
    SELECT * FROM ORDERLINE WHERE OL_W_ID = ? AND OL_D_ID = ? AND OL_O_ID < ? AND OL_O_ID > ?;
    SELECT COUNT(*) FROM STOCK WHERE S_I_ID = ? AND S_W_ID = ? AND S_QUANTITY < ?;
}

InferenceConstraints{
    OIC [W_ID]
    OIC [D_ID]
    OIC [C_ID]
    OIC [O_ID]
    OIC [I_ID]

    DIC [W_NAME]
    DIC [D_NAME]
    DIC [C_PHONE]
    DIC [I_NAME]
}

ConfidentialityConstraints{
    //Protect Stock
    //////////////////////////////

    // protect the amount of stored items
    [I_NAME, S_QUANTITY]
    [I_PRICE, S_QUANTITY]
    [I_DATA, S_QUANTITY]
    [I_IM_ID, S_QUANTITY]

    // protect the amount of ordered items
    [I_NAME, S_ORDER_CNT]
```

```
182        [I_PRICE, S_ORDER_CNT]
183        [I_DATA, S_ORDER_CNT]
184        [I_IM_ID, S_ORDER_CNT]
185        [I_NAME, S_REMOTE_CNT]
186        [I_PRICE, S_REMOTE_CNT]
187        [I_DATA, S_REMOTE_CNT]
188        [I_IM_ID, S_REMOTE_CNT]
189
190        // protect the amount of sold items
191        [I_NAME, S_YTD]
192        [I_PRICE, S_YTD]
193        [I_DATA, S_YTD]
194        [I_IM_ID, S_YTD]
195
196        // prevent the identifications of items via customer
              orders
197        [I_ID,OL_I_ID]
198
199        // protect attributes that contain confidential
              information
200        [H_DATA]
201        [O_ALL_LOCAL]
202        [OL_DIST_INFO]
203        [S_DATA]
204     }
205  }
```

For a detailed explanation of the data and query model as well as the meaning of each attribute we refer to [Tra10]. In the following, we provide the reasoning behind the specified ICs as well as a realistic protection objective which we achieve by specifying CCs.

The reasoning behind the ICs that are specified in lines 158-167 of the policy profile is as follows. The ID attribute values in the outsourced data (e.g., *W_ID,D_ID,...*) are randomly chosen and unique. Furthermore, their ordering bears no meaning. Their function is to act as primary or secondary keys which is why they are only relevant for the database and outside attackers can be assumed to have no background knowledge on them. As attackers can be assumed to have no background knowledge on the attribute values and their order bears no meaning, we argue that OICs can be specified for the ID attributes.

Furthermore, we specified DICs on the attributes *W_NAME, D_NAME, C_PHONE,* as well as *I_NAME.* Attribute values of those attributes can be considered unique. Thus, an attacker is not able to apply background knowledge on distinguishable ciphertexts to reveal the plaintext values.

Protection objective: In the following, we show which CCs have to be specified to prevent attackers from determining the sales strategy of the wholesale supplier. For instance, if based on the outsourced database an attacker observes that a large number of old items is stocked, she can infer that the supplier is forced to offer discounts in the near future. To address this protection objective, a risk manager can decide that it is too much of a risk that CSPs can determine for each item type ...

- ...the amount of stored items (*S_QUANTITY*),
- ...the amount of ordered items (*S_REMOTE_CNT*, *S_ORDER_CNT*),
- ...and the amount of sold items (*S_YTD*).

Thus, before the database is outsourced, it has to be enforced that the CSPs are not able to view how many items of a given item type are stored, ordered, and sold. In the following we show how CCs can be used to express this requirement.

A simple way would be to define the CCs `[S_QUANTITY]`, `[S_REMOTE_CNT]`, `[S_ORDER_CNT]`, `[S_YTD]`. Thus, the number or stored, ordered, and sold items are not accessible by the CSP. However, these CCs are stricter than necessary which potentially results in unnecessary efficiency overheads that are induced by CPIs. For instance, the amount of stored items *S_QUANTITY* is in fact not considered confidential but the mapping of the item to *S_QUANTITY* is.

An item is identified by its name (*I_NAME*). Thus, to protect the mapping between the name of the item and its quantity the CC `[I_NAME, S_QUANTITY]` can be specified. However, there are semantic dependencies of the attribute *I_NAME* on other attributes that have to be modeled by the user as they are not automatically covered by Securus. For instance, it can be possible to determine *I_NAME* based on the price of the item *I_PRICE*, the brand information of the item *I_DATA*, and the image ID of the item *I_IM_ID*. Thus, these attributes have to be considered equivalent to *I_NAME* and the CCs `[I_PRICE, S_QUANTITY]`, `[I_DATA, S_QUANTITY]`, and `[I_IM_ID, S_QUANTITY]` have to be specified (see lines 175-178 in the policy profile). CCs to protect the amount of ordered items (*S_REMOTE_CNT*, *S_ORDER_CNT*) and sold items (*S_YTD*) can be specified analogously (see lines 181-194 in the policy profile).

It is also possible that an attacker knows which customer ordered a particular item and use attribute values of the customer such as her name to identify the item. Furthermore, there is a semantic dependency between the number of sold items and the number of ordered items. Thus, if records of the ORDERLINE table can be linked to a specific item, the attacker can gain information on the number of ordered and sold items. Both semantic dependencies can be addressed by the CC `[I_ID, OL_I_ID]` (see line 197 in the policy profile). If either *I_ID* or *OL_ID* are not revealable by the attacker, orders in the orderline and therefore customers cannot be linked to a specific item.

Some attributes that contain information on the amount of stored, ordered, and sold items have to be protected as well. These attributes include *H_DATA*, *O_ALL_LOCAL*, *OL_DIST_INFO*, *S_DATA* (see lines 200-203 in the policy profile). For a detailed description of the attributes we refer to the TPCC specification [Tra10].

To keep the length of this thesis reasonable, we only focused on the objective of keeping the sales strategy of the wholesale supplier confidential. Other confidentiality objectives like the guaranteeing the protection of customer data may exist. These objectives can be mapped to CCs similarly to the objective of protecting the sales strategy of the wholesale supplier.

A.5 Adaptation to Non Query Observing Attackers

The version of Securus that we presented in Section 3.5 is able to enforce CCs against attackers that are able to observe queries on the data, including update, insert, and delete operations. If the purpose of Securus is to protect against attackers that compromise the CSP, copy the data, are discovered and locked out, it can be assumed that the attacker is not able to observe queries. If the assumption can be made that attackers are not able to observe queries, Securus can apply additional CPIs to enforce query requirements. Thus, potentially more efficient mediators are generated by Securus and a wider range of query requirements can be satisfied for a given set of confidentiality constraints.

The additional enforceable query requirements are shown in Table A.3 in boldface. For instance, the CPIs searchable encryption and encrypted B-Trees can be used to evaluate range and equality selections based on indistinguishable ciphertexts *only* if the attacker cannot observe queries (see Section 3.4.3). The modified CPI selection table can be used in step 4 of Securus' policy transformation process (see Figure 3.8 in Section 3.5). As the modified CPI selection table allows to evaluate query conditions such as equality selections on indistinguishable ciphertexts, Securus' ILP problem formalization has to be adapted. This can be easily achieved. For instance, the fact that it is possible to evaluate equality selections based on indistinguishable ciphertexts can be included in the ILP problem by altering the ILP constraint in Equation A.3 as follows:

$$\forall j \in S, \forall q \in QP, \forall i \in e(q) : P_{i,j} + O_{i,j} + D_{i,j} + \boldsymbol{I_{i,j}} - Q_{q,j} \geq 0$$

A.6 Extensions to the Target Function

To determine the optimal set of CPIs, the version of Securus that we presented in Section 3.5 only considered the query latency and transmission overhead that is induced by CPIs. We argue that storage overhead is a less important optimization objective in most deployment scenarios, as reasonable query latency is paramount for most database applications and network transmission costs are dominating storage costs in most cloud scenarios[2]. In many database scenarios, data is much more frequently queried than stored. Thus, the amount of data that is transmitted over the network typically exceeds by far the amount of storage space that is needed to store the data. For instance, the identity database stored at the KIT contains roughly 45000 records and has a size of around 2 GB. Maintenance jobs that query the database to achieve consistency and services that retrieve identity data for each user that logs in incur a transmission overhead that easily exceeds 2 TB within a single month.

Furthermore, the worst case storage overhead that is induced by Securus is bound by a fixed factor. In the worst case each SP has to store ciphertexts of each available CPI. Since the storage overhead is bounded and storage space is considered cheap compared to network capacities and query latency overhead, we feel that the storage overhead

[2] For instance, in Amazons S3 storage service, storing one GB of data costs 0.03$ per month and downloading one GB of data from S3 costs 0.09$: http://aws.amazon.com/de/s3/pricing/

		Attribute representation from a CSP's perspective			
Type	Query component	Plaintext	Order-preserving ciphertexts	Distinguishable ciphertexts	Indistinguishable ciphertexts
1	Attr1 = ?	✓	✓ e.g., OPE [BCLO09]	✓ e.g., keyed hashes	✓ **e.g., enc. B-tree [DCdVFP+11]**
2	Attr1 < ?	✓	✓ e.g., OPE [BCLO09]	✓ **e.g., enc. B-tree [DCdVFP+11]**	✓ **e.g., enc. B-tree [DCdVFP+11]**
3	Attr1 (NOT) LIKE ?	✓	✓ e.g., SE schemes [SWP00]	✓ **e.g., SE schemes [SWP00]**	✓ **e.g., SE schemes [SWP00]**
4	Attr1 IN ?	✓	✓ e.g., OPE [BCLO09]	✓ e.g., keyed hashes	✓ **e.g., enc. B-tree [DCdVFP+11]**
5	SUM(Attr1)	✓	✓ e.g., Paillier [Pai99]	✓ e.g., Paillier [Pai99]	✓ e.g., Paillier [Pai99]
6	SUM(Attr1 * Attr2)	✓			
7	COUNT (DISTINCT Attr1)	✓	✓ e.g., OPE [BCLO09]	✓ e.g., keyed hashes	
8	GROUP BY Attr1	✓	✓ e.g., OPE [BCLO09]	✓ e.g., keyed hashes	
9	SORTED BY Attr1	✓	✓ e.g., OPE [BCLO09]		
10	SUBSTRING(Attr1)	✓			
11	HAVING SUM(Attr1) < ?	✓			
...	...	...	...	...	...

Table A.3: Attribute ciphertext representations required to efficiently evaluate various query components. The query requirements that are enforceable due to the assumption that attackers cannot observe queries are highlighted in boldface.

is not an issue in most cases and, therefore, optimizing for it is not worthwhile in many deployment scenarios. Nevertheless, we provide an outlook on how the storage overhead that is induced by CPIs can be accounted for in the ILP problem formalization of Securus.

Storage costs: Let s_i^c be the storage cost in bytes that is induced for storing representations of attribute i that are created by applying the CPI c. The storage cost s_i^c that is induced for storing an attribute i has to be determined for each CPI c. For instance, in case of AES encryption, the storage cost s_i^{AES} amounts to the size of the ciphertexts that result from encrypting plaintext records of i. Furthermore, binary variables $y_{c,j,i}$ state that CSP j stores representations of attribute i that are created by CPI c. In the ILP problem, $y_{c,j,i}$ are free variables while s_i^c are constants.

As shown in Equation A.23 the storage overhead can be determined by summing up the storage costs that are induced by the applied CPIs at all CSPs:

$$\sum_{i \in A} \sum_{j \in S} \sum_{c \in CPIs} (s_i^c \cdot y_{c,j,i}) \tag{A.23}$$

The allocations of the free variables $y_{c,j,i}$ are subject to constraints. The constraints depend on the CPI type c that $y_{c,j,i}$ addresses. In the following we exemplarily list the constraints for plaintext values and the keyed hashes CPI.

Plaintext ($c = 1$): The CSP j has to store plaintext values of an attribute i ($y_{1,j,i} = 1$) if a query q exists for which i is relevant ($i \in q$) and the CSP j can evaluate the query q ($Q_{q,j} = 1$) and may store plaintext values for i ($P_{i,j} = 1$). This condition is captured by the ILP constraint that is shown in Equation A.24:

$$\forall q \in QP, \forall i \in q, \forall j \in S: \quad y_{1,j,i} - Q_{q,j} - P_{i,j} \quad \geq -1 \tag{A.24}$$

Keyed hashes ($c = 2$): The CSP j has to store keyed hashes of an attribute i ($y_{2,j,i} = 1$) if a query q exists for which i is relevant and for which the keyed hashes CPI is applied. Securus applies the keyed hashes CPI on an attribute i at CSP j if CSP j may only store order-preserving ($O_{i,j} = 1$) or distinguishable ciphertexts ($D_{i,j} = 1$) of the attribute and can evaluate a query q ($Q_{q,j} = 1$) that contains conditions of the types 1,4,7, or 8 (see Table 3.5). These conditions are mapped to the ILP constraints that are shown in Equation A.25-A.26:

$$\forall q \in QP, \forall i \in q(\{1,4,7,8\}), \forall j \in S: \quad y_{2,j,i} - Q_{q,j} - O_{i,j} \quad \geq -1 \tag{A.25}$$
$$\forall q \in QP, \forall i \in q(\{1,4,7,8\}), \forall j \in S: \quad y_{2,j,i} - Q_{q,j} - D_{i,j} \quad \geq -1 \tag{A.26}$$

ILP Constraints on $y_{c,j,i}$ for other CPIs can be derived analogously to those already presented.

Taking storage overhead into account results in a bigger ILP problem with $|C| \cdot |S| \cdot |A|$ additional free variables. An investigation of whether the ILP formulation is still solvable in acceptable time is beyond the scope of this thesis and left for future work. In particular, the initial ideas to reduce the ILP's problem size which we provided in Section 3.5.7 can be followed up on to address the problem.

A.7 Extension via Anonymity Constraints

In this section, on the example of ℓ-diversity requirements, we provide a first step towards integrating anonymity requirements in Securus. We show how the ILP problem of Securus can be extended to support ℓ-diversity requirements. Thus, when Securus generates the mediator, an optimal choice can be made on whether to evaluate queries on ℓ-diversified index tables as proposed in Section 3.6 or to apply CPIs instead.

The ℓ-diversity policy $l=$ `[3:A,B:B,C]` denotes that 3-diversity is required. The QID attributes are A and B. The attributes B and C are considered sensitive. The power set $2^l = \{\{A,B\},\{A,C\},\{A,B,C\},\{B,C\}\}$ contains all possible attribute combinations that would require to ℓ-diversify the attribute values of the contained attributes if they can be viewed by the CSP.

The optimization problem that amounts from jointly considering CPIs and ℓ-diversification of the outsourced data can be summarized by the following question. Based on a query that requires to evaluate conditions on an attribute $i \in x$ that is part of an ℓ-diversity policy, when is it more efficient to apply a CPI to attribute i than evaluating queries on pre-joined, ℓ-diversified index tables that contain attribute i? For the sake of simplicity, in the following, we assume that all attributes

$i \in x$ that are relevant to evaluate a query condition either have to be protected via CPIs or have to be anonymized. We leave the problem of optimally choosing which query conditions should be evaluated at the mediator rather than based on CPIs or anonymized data at the CSP for future work.

Securus' ILP problem formalization can be extended to support anonymity requirements as follows.

Additional constants:

l : The ℓ-diversity constraint.

$c^l_{q,x}$: Latency costs of evaluating query q based on a pre-joined, ℓ-diversified index table that contains the attributes x. In particular, Dividat efficiency models can be used to determine these costs.

$t^l_{q,x}$: Transmission costs of evaluating query q based on a pre-joined, ℓ-diversified index table that contains the attributes x. In particular, Dividat efficiency models can be used to determine these costs.

Additional free variables:

$a_{x,q}$: Query q is evaluated based on an pre-joined, ℓ-diversified index table that contains the attributes x (in this case, $a_{x,q} = 1$, otherwise, $a_{x,q} = 0$). Other attributes that are relevant for the query are assumed to be protected via CPIs.

Target function additions:

The following function can be added to the existing target function of Securus to account for the costs that are induced by evaluating queries based on ℓ-diversified, pre-joined index tables.

$$\sum_{q \in QP} \sum_{x \in 2^l} (c^l_{q,x} \cdot a_{x,q} + t^l_{q,x} \cdot a_{x,q}) \tag{A.27}$$

The query latency and transmission costs of outsourcing anonymized plaintext values of attributes in the set x to a CSP depend on which queries are evaluated on these outsourced values. The query latency cost of evaluating query q on a pre-joined, ℓ-diversified index table that contains the attributes x is denoted by $c^l_{q,x}$. Thus, as shown in the target function addition in Equation A.27, the query latency costs that are induced by evaluating queries on pre-joined index tables can be calculated by $\sum_{q \in QP} \sum_{x \in 2^l} c^l_{q,x} \cdot a_{x,q}$. The transmission costs can be calculated analogously by $\sum_{q \in QP} \sum_{x \in 2^l} t^l_{q,x} \cdot a_{x,q}$.

Additional ILP constraints:

When does a query q have to be evaluated on a pre-joined, ℓ-diversified index table that contains the attributes x, i.e., when does $a_{x,q} = 1$ hold?

If a CSP is responsible to evaluate a query q, but may only store plaintext values of the attributes $x \in 2^l$, Securus has to outsource ℓ-diversified index tables to satisfy the ℓ-diversity requirement l. Thus, attribute combinations that require ℓ-diversification can be outsourced as plaintexts but have to be anonymized first. Evaluating a query on a pre-joined index table incurs a cost that has to be captured in the ILP problem. To calculate the induced cost of evaluating queries on ℓ-diversified, pre-joined index tables and compare it to the cost of applying CPIs, it has to be captured which queries have to be evaluate on which pre-joined index tables. This is achieved by the free variables $a_{x,q}$. Notice that the approach to penalize CPI combinations that require to evaluate queries on anonymized index tables with a higher cost implies that no additional ILP constraints are necessary to enforce ℓ-diversity requirements. Likewise, the existing constraints to enforce QPs do not have to be altered as ℓ-diversity requirements do not prevent from outsourcing attribute values in plaintext.

The query q has to be evaluated based on an index table that contains the attributes x ($a_{x,q} = 1$) if

1. the CSP j that is responsible to evaluate the query q ($Q_{q,j} = 1$) can reveal all attributes $i \in x$, i.e., no CPIs are used to protect the attributes. The term $\sum_{i \in x} Z_{i,j} - |x| + Q_{q,j} - 1$ equals 0 if this is the case. If either CSP j is not responsible to evaluate the query q ($Q_{q,j} - 1 < 0$) or not all attributes $i \in x$ can be revealed by CSP j ($\sum_{i \in x} Z_{i,j} - |x| < 0$), the term is smaller than 0.

2. the query q is not evaluated on another anonymized index table that contains the attributes $x' \in 2^l$. If this is the case, the term $- \sum_{x' \in 2^l} a_{x',q}$ equals 0, otherwise it is smaller than 0.

Combining both conditions, the ILP constraint on when $a_{x,q} = 1$ has to hold can be formalized as shown in Equation A.28:

$$\forall x \in 2^l, \forall q \in QP, \forall j \in S \ : \ \sum_{i \in x} Z_{i,j} - |x| + Q_{q,j} - 1 - \sum_{x' \in 2^l \setminus \{x\}} a_{x',q} < a_{x,q} \quad \text{(A.28)}$$

The constraint enforces that $a_{x,q} = 1$ holds if both conditions 1) and 2) hold, i.e., $\sum_{i \in x} Z_{i,j} - |x| + Q_{q,j} - 1 = 0$ and $- \sum_{x' \in 2^l \setminus \{x\}} a_{x',q} = 0$. If this is not the case, the free variable $a_{x,q}$ can take both values 1 and 0.

In some cases $a_{x,q}$ must not equal 1. For $a_{x,q} = 1$, an index table that contains ℓ-diversified plaintext values has to be outsourced to one CSP. The outsourcing of plaintext values, however, can contradict with confidentiality constraints. It has to hold that the CSP j that is responsible to evaluate query q ($Q_{q,j} = 1$) is allowed to reveal plaintext values of all attributes $i \in x$ of the index table. This can be ensured by the ILP constraint shown in Equation A.29:

$$\forall x \in 2^l, \forall q \in QP, \forall j \in S \ : \ (1 - Q_{q,j}) + \frac{1}{|x|} \sum_{i \in x} Z_{i,j} \geq a_{x,q} \quad \text{(A.29)}$$

If CSP j is not responsible to evaluate the query q ($Q_{q,j} = 0$), the term $(1 - Q_{q,j}) + \frac{1}{|x|} \sum_{i \in x} Z_{i,j}$ is bound to be bigger or equal to 1. This is due to the fact that $1 \geq \frac{1}{|x|} \sum_{i \in x} Z_{i,j} \geq 0$ holds in any case. Thus, if $Q_{q,j} = 0$ holds, the constraint has no influence on $a_{x,q}$ as $a_{x,q}$ can still take both values 0 and 1. If the CSP j is responsible ($Q_{q,j} = 1$) and may reveal plaintext values of all attributes $i \in x$ contained the index table, it holds that $\frac{1}{|x|} \sum_{i \in x} Z_{i,j} = 1$ and the term equals 1. Thus, $a_{x,q}$ may take the value 1. If the CSP j is responsible, but must not reveal plaintext values of an attribute $i \in x$ that is contained the index table, i.e., $\frac{1}{|x|} \sum_{i \in x} Z_{i,j} < 1$ and $1 - Q_{q,j} = 0$, the term is smaller to 1. In this case, the constraint enforces that $a_{x,q}$ must equal 0.

Further evaluation of the feasibility of the proposed extension to integrate ℓ-diversity requirements with Securus in practice is beyond the scope of this thesis. However, the extension shows that the concepts of anonymized DaaS and confidential DaaS can be jointly addressed by a single framework and synergies can be leveraged. In particular, the proposed concept enforces ℓ-diversity requirements via CPIs if it can be considered more efficient than applying anonymization techniques. The Securus extension can even satisfy ℓ-diversity requirements based on a mix of CPIs and anonymization techniques if it is beneficial. Furthermore, via this extension Securus can provide ℓ-diversity guarantees for attribute values on which complex queries such as regular expressions have to be evaluated. This was not possible by relying on CPIs alone.

B
Appendix: Dividat

B.1 Formalization of the Anatomized, ℓ-Diversified Index Optimization Problem

Finding a database indexing strategy that is tuned to satisfy the anonymity requirements of a deployment scenario and optimizes efficiency at the same time constitutes a challenging optimization problem that is hard to grasp. In Section 3.6 we proposed Dividat, a framework to build efficiency models for anonymized indexes and applied it to build models for the case of anatomized, ℓ-diversified databases. We showed how these models can be used to determine which index table should be used to execute a query most efficiently. In this section, based on the proposed efficiency models we formalize the problem of finding an efficiency-optimal set of ℓ-diversified index tables for a given query workload.

A formal definition of the optimization problem to determine the optimal set of index tables for a query workload Q is given in the following.

Constants:

ℓ	: ℓ-diversity parameter.
$QidSens$	: Set of sensitive attributes that are contained in the QID.
$SensOnly$	: Set of sensitive attributes that are not contained in the QID.
A	: Set of attributes
$\mathcal{I}$	: Set of candidate index tables $\{1, \ldots, n\}$ (fixed to a certain number n).
$\mathcal{Q}$	: Set of queries that are contained in the workload.
N_M	: Number of records that are contained the outsourced database.
q_a	: Set of attribute values that match query q's condition for attribute a.
V_a	: Domain of outsourced values for attribute a.
s_a	: Size of attribute a's values.
a_q	: Transmission overhead that is incurred by transmitting one result record of query q.
w_s	: Weight for storage costs (user preference).
w_t	: Weight for transmission costs (user preference).
w_l	: Weight for query latency costs (user preference).
b, c, f	: Parameters of the query latency model.

Free variables:

S_I	: Storage costs that are induced by index table I.
$T_{I,q}$	: Transmission overhead of executing query q on index table I.
$L_{I,q}$	: Latency overhead of executing query q on index table I.
c_q	: Weighted latency and transmission cost for executing query q.
$i_{a,I}$	: Attribute a is contained in index table I ($i_{a,I} = 1$).
z_I	: Index table I contains a *SensOnly* attribute ($z_I = 1$).
N_I	: Number of records that are contained in index table I.
$e_{q,I}$	: Query q is planned to be executed based on index table I ($e_{q,I} = 1$).
$t_{I,q}$	: Expected number of transmitted records when executing query q on index table I.
$r_{I,q}$	: Expected number of records that match query q in index table I.

Optimization problem:

$$\min w_s \cdot \textstyle\sum_{I \in \mathcal{I}} S_I + \frac{1}{|\mathcal{Q}|} \sum_{q \in \mathcal{Q}} c_q \tag{B.1}$$

s.t.

$$\forall I \in \mathcal{I}: \quad \textstyle\sum_{a \in A} (s_a \cdot i_{a,I}) \cdot N_I = S_I \tag{B.2}$$

$$\forall I \in \mathcal{I}: \quad \ell^{\sum_{a \in QidSens} i_{a,I} + z_I} \cdot N_M = N_I \tag{B.3}$$

$$\forall I \in \mathcal{I}: \quad \textstyle\sum_{a \in SensOnly} i_{a,I} \leq z_I \cdot |A| \tag{B.4}$$

$$\forall q \in \mathcal{Q}: \quad \textstyle\sum_{I \in \mathcal{I}} e_{q,I} \cdot (w_l \cdot L_{I,q} + w_t \cdot T_{I,q}) = c_q \tag{B.5}$$

$$\forall q \in \mathcal{Q}: \quad \textstyle\sum_{I \in \mathcal{I}} e_{q,I} > 0 \tag{B.6}$$

$$\forall I \in \mathcal{I} : \forall q \in \mathcal{Q}: \quad a_q \cdot t_{I,q} = T_{I,q} \tag{B.7}$$

$$\forall I \subset \mathcal{I} : \forall q \subset \mathcal{Q}: \quad b \cdot t_{I,q} + c \cdot r_{I,q} + f = L_{I,q} \tag{B.8}$$

$$\forall I \in \mathcal{I} : \forall q \in \mathcal{Q}: \quad \textstyle\prod_{a \in A} \left(\frac{|q_a|}{|V_a|}\right)^{i_{a,I}} \cdot N_I = r_{I,q} \tag{B.9}$$

$$\forall I \in \mathcal{I} : \forall q \in \mathcal{Q} : \textstyle\prod_{a \in SensOnly \cup QidSens} \left(1 - \prod_{i=0}^{\ell-1} \left(1 - \frac{|q_a|}{|V_a| - i}\right)\right)^{i_{a,I}} \cdot \frac{N_M}{\ell} \cdot \ell \cdot \prod_{a \in A/(SensOnly \cup QidSens)} \left(\frac{|q_a|}{|V_a|}\right)^{i_{a,I}} = t_{I,q} \tag{B.10}$$

$$\forall a \in A : \forall I \in \mathcal{I}: \quad i_{a,I} \in \{0, 1\} \tag{B.11}$$

$$\forall I \in \mathcal{I}: \quad z_I \in \{0, 1\} \tag{B.12}$$

$$\forall I \in \mathcal{I} : \forall q \in \mathcal{Q}: \quad e_{q,I} \in \{0, 1\} \tag{B.13}$$

The shown optimization problem determines for each candidate index table $I \in \{1, \ldots, n\}$ the attributes that have to be contained in the index table to minimize the overhead function $o(\mathcal{I}, \mathcal{Q})$ for a query workload $\mathcal{Q}$. As the maximum number of index tables is fixed to n, the allocation of the free variables $i_{a,I}$ states whether attribute a is contained in index table I and, thus, constitutes the solution space.

The optimization goal is to minimize the overhead function $o(\mathcal{I}, \mathcal{Q})$ that considers the weighted storage overhead S_I for each index table I as well as the weighted query latency and transmission overhead c_q for each query q (Equation B.1). The storage overhead that is induced by an index table I can be calculated based on the SO-model (see Equation 3.11), i.e., by multiplying the size of each contained record with the number of records N_I in the index table (Equation B.2). For the ℓ-diversified database outsourcing as proposed in Section 3.6.2 the number of records N_I that are contained in the index table can be calculated based on the introduced NR-model (Equation 3.14) which is expressed by the ILP constraint shown in Equation B.3. To model the $\delta(x)$ function, we introduced a helping variable z_I which equals 1 if a sensitive attribute that is not part of the QID is contained in the index table and 0 otherwise. This is ensured by the ILP constraint shown in Equation B.4.

The weighted query latency and transmission overhead c_q for each query q is determined by the index table I based on which query q is executed ($e_{q,I} = 1$). This is expressed by the ILP constraint shown in Equation B.5. Each query has to be executed based on at least one index table (Equation B.6). The transmission overhead $T_{I,q}$ and the latency overhead $L_{I,q}$ that is induced when evaluating query q on index table I can be calculated by the generic TO- and QL-model that we proposed in Section

3.6.3. The number of records $r_{I,q}$ that match a query q in the index table I and the number of records $t_{I,q}$ that have to be transmitted to the mediator can be estimated based on the MR- and TR-models we proposed for the case of ℓ-diversified database outsourcing in Section 3.6.4. These models are cast into the ILP constraints that are shown in Equation B.9 and Equation B.10.

The optimization problem can be solved for a given query workload, if the parameterization of the SO-, TO-, and the QL-model is known. Determining the parameterization of the SO-model is simple, as the storage space that is needed to store a single record in an index table can be easily calculated by $\sum_{a \in A}(s_a \cdot i_{a,I})$, i.e., by adding up the sizes of all attributes that are contained in the index table I. The parameterization of the TO-model is also easy, as the overhead for transmitting a single query results record a_q can be analytically determined for each query q. Getting hold of the parameterization b, c, and f of the QL-model can be complicated as the parameters can depend on how the database management system stores and uses the index tables. In particular, based on some database management system the parameters may vary for different index tables. Thus, it has to be determined whether they remain constant for different index tables. This strongly depends on the database management system that the CSP operates and the utilized anonymization technique. If the factors are not constant, it can be tried to develop separate models for them. If this is not possible, each anonymized index table candidate has to be built and the parameters have to be determined based on performance measurements as shown in Section 3.6.3.

The optimization problem of building indexes for anatomized, ℓ-diversified database outsourcing was hard to determine intuitively. We were able to clearly formulate this optimization problem based on the efficiency models that we developed based on Dividat. Proposing and evaluating methods to solve this optimization problem is beyond the scope of this thesis.

Bibliography

[ABG$^+$05] G. Aggarwal, M. Bawa, P. Ganesan, H. Garcia-Molina, K. Kenthapadi, R. Motwani, U. Srivastava, D. Thomas, and Y. Xu. Two can keep a secret: A distributed architecture for secure database services. In *Proceedings of the Second Conference on Innovative Data Systems Research*, 2005.

[ABR$^+$10] P. Angin, B. Bhargava, R. Ranchal, N. Singh, M. Linderman, L. Ben Othmane, and L. Lilien. An entity-centric approach for privacy and identity management in cloud computing. In *Proceedings of the 29th IEEE Symposium on Reliable Distributed Systems*, pages 177–183, 2010.

[ACN06] S. Agrawal, E. Chu, and V. Narasayya. Automatic physical design tuning: Workload as a sequence. In *Proceedings of the ACM International Conference on Management of Data*, SIGMOD '06, pages 683–694, New York, NY, USA, 2006. ACM.

[AFG$^+$09] M. Armbrust, A. Fox, R. Griffith, A. Joseph, R. Katz, A. Konwinski, G. Lee, D. Patterson, A. Rabkin, I. Stoica, and M. Zaharia. Above the clouds: A Berkeley view of cloud computing. Technical report, Berkeley, 2009.

[AGH11] D. Achenbach, M. Gabel, and M. Huber. Mimosecco: A middleware for secure cloud storage. In D. D. Frey, S. Fukuda, and G. Rock, editors, *Improving Complex Systems Today*, Advanced Concurrent Engineering, pages 175–181. Springer London, 2011.

[AIM10] L. Atzori, A. Iera, and G. Morabito. The internet of things: A survey. *Computer Networks*, 54(15):2787–2805, October 2010.

[AKSX04] R. Agrawal, J. Kiernan, R. Srikant, and Y. Xu. Order preserving encryption for numeric data. In *Proceedings of the ACM International Conference on Management of Data*, SIGMOD '04, pages 563–574, New York, NY, USA, 2004. ACM.

[AMG07] C. Aguilar-Melchor and P. Gaborit. A lattice-based computationally-efficient private information retrieval protocol. In *Western European Workshop on Research in Cryptology (WEWoRC'2007), Bochum, Germany. Book of Abstracts*, pages 50–54, 2007.

[Apt03] K. Apt. *Principles of constraint programming.* Cambridge University Press, 2003.

[Avr03] M. Avriel. *Nonlinear programming: analysis and methods.* Courier Dover Publications, 2003.

[AY08] C. C. Aggarwal and P. S. Yu. A framework for condensation-based anonymization of string data. *Data Mining and Knowledge Discovery*, 16(3):251–275, 2008.

[BA07] R. Boutaba and I. Aib. Policy-based management: A historical perspective. *Journal of Network and Systems Management*, 15(4):447–480, 2007.

[BBO07] M. Bellare, A. Boldyreva, and A. O'Neill. Deterministic and efficiently searchable encryption. In A. Menezes, editor, *Advances in Cryptology - CRYPTO 2007*, volume 4622 of *Lecture Notes in Computer Science*, pages 535–552. Springer, Berlin Heidelberg, 2007.

[BC05] N. Bruno and S. Chaudhuri. Automatic physical database tuning: A relaxation-based approach. In *Proceedings of the ACM International Conference on Management of Data*, SIGMOD '05, pages 227–238, New York, NY, USA, 2005. ACM.

[BCLO09] A. Boldyreva, N. Chenette, Y. Lee, and A. O'Neill. Order-preserving symmetric encryption. In A. Joux, editor, *Advances in Cryptology - EUROCRYPT 2009*, volume 5479 of *Lecture Notes in Computer Science*, pages 224–241. Springer, Berlin Heidelberg, 2009.

[Bei11] A. Beimel. Secret-sharing schemes: A survey. In Y. Chee, Z. Guo, S. Ling, F. Shao, Y. Tang, H. Wang, and C. Xing, editors, *Coding and Cryptology*, volume 6639 of *Lecture Notes in Computer Science*, pages 11–46. Springer Berlin Heidelberg, 2011.

[BHM77] S. Bradley, A. Hax, and T. Magnanti. *Applied mathematical programming.* Addison Wesley, 1977.

[BI93] M. Bertilsson and I. Ingemarsson. A construction of practical secret sharing schemes using linear block codes. In J. Seberry and Y. Zheng, editors, *Advances in Cryptology — AUSCRYPT '92*, volume 718 of *Lecture Notes in Computer Science*, pages 67–79. Springer Berlin Heidelberg, 1993.

[BIKR02] A. Beimel, Y. Ishai, E. Kushilevitz, and J.-F. Raymond. Breaking the O(n1(2k-1)) barrier for information-theoretic private information retrieval. In *Foundations of Computer Science, 2002. Proceedings. The 43rd Annual IEEE Symposium on*, pages 261–270, 2002.

[BJO09] K. D. Bowers, A. Juels, and A. Oprea. Proofs of retrievability: Theory and implementation. In *Proceedings of the 2009 ACM Workshop on Cloud Computing Security*, CCSW '09, pages 43–54, New York, NY, USA, 2009. ACM.

[BKN02] M. Bellare, T. Kohno, and C. Namprempre. Authenticated encryption in SSH: Provably fixing the SSH binary packet protocol. In *Proceedings of the 9th ACM Conference on Computer and Communications Security*, CCS '02, pages 1–11, New York, NY, USA, 2002. ACM.

[BM92] S. Bellovin and M. Merritt. Encrypted key exchange: password-based protocols secure against dictionary attacks. In *Proceedings of the IEEE Computer Society Symposium on Research in Security and Privacy*, pages 72–84, May 1992.

[BM93] S. M. Bellovin and M. Merritt. Augmented encrypted key exchange: A password-based protocol secure against dictionary attacks and password file compromise. In *Proceedings of the 1st ACM Conference on Computer and Communications Security*, CCS '93, pages 244–250, New York, NY, USA, 1993. ACM.

[BP14] N. Bjørner and A.-D. Phan. *νz*-maximal satisfaction with z3. In *Proceedings of International Symposium on Symbolic Computation in Software Science (SCSS)*, 2014.

[BPFS09] E. Bertino, F. Paci, R. Ferrini, and N. Shang. Privacy-preserving digital identity management for cloud computing. *IEEE Data Engineering Bulletin*, 32(1):21–27, 2009.

[BS03] A. Beimel and Y. Stahl. Robust information-theoretic private information retrieval. In *Security in Communication Networks*, pages 326–341. Springer, Berlin Heidelberg, 2003.

[BS11] S. Bajaj and R. Sion. TrustedDB: A trusted hardware based database with privacy and data confidentiality. In *Proceedings of the ACM International Conference on Management of Data*, SIGMOD '11, pages 205–216, New York, NY, USA, 2011. ACM.

[BV04] S. P. Boyd and L. Vandenberghe. *Convex optimization*. Cambridge university press, 2004.

[BV09] S. Boyd and L. Vandenberghe. *Convex optimization*. Cambridge university press, 2009.

[Can01] R. Canetti. Universally composable security: a new paradigm for cryptographic protocols. In *Proceedings of the 42nd IEEE Symposium on the Foundations of Computer Science*, pages 136–145, Oct 2001.

[CCG+99] S. Chaudhuri, E. Christensen, G. Graefe, V. R. Narasayya, and M. J. Zwilling. Self-tuning technology in Microsoft SQL server. *IEEE Data Engineering Bulletin*, 22(2):20–26, 1999.

[CDL+13] J. Camenisch, M. Dubovitskaya, A. Lehmann, G. Neven, C. Paquin, and F.-S. Preiss. Concepts and languages for privacy-preserving attribute-based authentication. In S. Fischer-Hübner, E. Leeuw, and C. Mitchell, editors, *Policies and Research in Identity Management*, volume 396 of *IFIP Advances in Information and Communication Technology*, pages 34–52. Springer, Berlin Heidelberg, 2013.

[CDV+05] A. Ceselli, E. Damiani, S. D. C. D. Vimercati, S. Jajodia, S. Paraboschi, and P. Samarati. Modeling and assessing inference exposure in encrypted databases. *ACM Transactions on Information System Security*, 8(1):119–152, February 2005.

[CdVFP+13] S. Capitani di Vimercati, S. Foresti, S. Paraboschi, G. Pelosi, and P. Samarati. Distributed shuffling for preserving access confidentiality. In J. Crampton, S. Jajodia, and K. Mayes, editors, *Computer Security – ESORICS 2013*, volume 8134 of *Lecture Notes in Computer Science*, pages 628–645. Springer, Berlin Heidelberg, 2013.

[CG97] B. Chor and N. Gilboa. Computationally private information retrieval (extended abstract). In *Proceedings of the twenty-ninth annual ACM Symposium on Theory of Computing*, STOC '97, pages 304–313, New York, NY, USA, 1997. ACM.

[Cha98] S. Chaudhuri. An overview of query optimization in relational systems. In *Proceedings of the 17th ACM Symposium on Principles of Database Systems (PODS)*, pages 34–43, 1998.

[Che03] H. W. Chesbrough. *Open innovation: The new imperative for creating and profiting from technology*. Harvard Business Press, 2003.

[CI09] D. Chadwick and G. Inman. Attribute aggregation in federated identity management. *IEEE Computer*, 42(5):33 –40, 2009.

[CKGS98] B. Chor, E. Kushilevitz, O. Goldreich, and M. Sudan. Private information retrieval. *Journal of the ACM*, 45(6):965–981, November 1998.

[CKS10] J. Camenisch, M. Kohlweiss, and C. Soriente. Solving revocation with efficient update of anonymous credentials. In J. Garay and R. De Prisco, editors, *Security and Cryptography for Networks*, volume 6280 of *Lecture Notes in Computer Science*, pages 454–471. Springer Berlin Heidelberg, 2010.

[CL02] J. Camenisch and A. Lysyanskaya. Dynamic accumulators and application to efficient revocation of anonymous credentials. In M. Yung, editor, *Advances in Cryptology — CRYPTO 2002*, volume 2442 of *Lecture Notes in Computer Science*, pages 61–76. Springer Berlin Heidelberg, 2002.

[CM05] Y.-C. Chang and M. Mitzenmacher. Privacy preserving keyword searches on remote encrypted data. In J. Ioannidis, A. Keromytis, and M. Yung, editors, *Applied Cryptography and Network Security*, volume 3531 of *Lecture Notes in Computer Science*, pages 442–455. Springer, Berlin Heidelberg, 2005.

[CN97] S. Chaudhuri and V. R. Narasayya. An efficient, cost-driven index selection tool for Microsoft SQL server. In *VLDB*, volume 97, pages 146–155, 1997.

[CN07] S. Chaudhuri and V. Narasayya. Self-tuning database systems: A decade of progress. In *Proceedings of the 33rd International Conference on Very Large Data Bases*, VLDB '07, pages 3–14. VLDB Endowment, 2007.

[Coo71] S. A. Cook. The complexity of theorem-proving procedures. In *Proceedings of the third annual ACM Symposium on Theory of Computing*, STOC '71, pages 151–158. ACM, 1971.

[Dam02] N. C. Damianou. *A policy framework for management of distributed systems*. PhD thesis, Imperial College, 2002.

[DCdVFJ+14] S. De Capitani di Vimercati, S. Foresti, S. Jajodia, G. Livraga, S. Paraboschi, and P. Samarati. Fragmentation in presence of data dependencies. *IEEE Transactions on Dependable and Secure Computing*, 11(6):510–523, Nov 2014.

[DCdVFLS11] S. De Capitani di Vimercati, S. Foresti, G. Livraga, and P. Samarati. Protecting privacy in data release. In A. Aldini and R. Gorrieri, editors, *Foundations of Security Analysis and Design VI*, volume 6858 of *Lecture Notes in Computer Science*, pages 1–34. Springer Berlin Heidelberg, 2011.

[DCdVFP+11] S. De Capitani di Vimercati, S. Foresti, S. Paraboschi, G. Pelosi, and P. Samarati. Efficient and private access to outsourced data. In *Proceedings of the 31st International Conference on Distributed Computing Systems*, ICDCS '11, pages 710–719, Washington, DC, USA, 2011. IEEE Computer Society.

[DDJ+03] E. Damiani, S. De Capitani di Vimercati, S. Jajodia, S. Paraboschi, and P. Samarati. Balancing confidentiality and efficiency in untrusted relational DBMSs. In *Proceedings of the ACM conference on Computer and Communications Security (CCS)*, 2003.

[DDLS01] N. Damianou, N. Dulay, E. Lupu, and M. Sloman. The Ponder policy specification language. In M. Sloman, E. Lupu, and J. Lobo, editors, *Policies for Distributed Systems and Networks*, volume 1995 of *Lecture Notes in Computer Science*, pages 18–38. Springer Berlin Heidelberg, 2001.

[DdVF+05] E. Damiani, S. D. C. di Vimercati, S. Foresti, S. Jajodia, S. Paraboschi, and P. Samarati. Key management for multi-user encrypted databases. In *Proceedings of the ACM Workshop on Storage Security and Survivability*, StorageSS '05, pages 74–83, New York, NY, USA, 2005. ACM.

[Deg15] A. Degitz. Design and performance analysis of database-as-a-service protocols with access pattern obfuscation. Master's thesis, KIT, 2015.

[DFJ+13a] S. De Capitani di Vimercati, S. Foresti, S. Jajodia, G. Livraga, S. Paraboschi, and P. Samarati. Extending loose associations to multiple fragments. In *Proceedings of the 27th Annual IFIP WG 11.3 Working Conference on Data and Applications Security and Privacy (DBSec)*, July 2013.

[DFJ+13b] S. De Capitani di Vimercati, S. Foresti, S. Jajodia, S. Paraboschi, and P. Samarati. On information leakage by indexes over data fragments. In *Proceedings of the 1st International Workshop on Privacy-Preserving Data Publication and Analysis (PrivDB)*, April 2013.

[DGMS01] P. T. Devanbu, M. Gertz, C. U. Martel, and S. G. Stubblebine. Authentic third-party data publication. In *Proceedings of the IFIP TC11/ WG11.3 Fourteenth Annual Working Conference on Database Security: Data and Application Security, Development and Directions*, pages 101–112, 2001.

[DMB11] L. De Moura and N. Bjørner. Satisfiability modulo theories: Introduction and applications. *Communications of the ACM*, 54(9):69–77, September 2011.

[Dwo06] C. Dwork. Differential privacy. In *Automata, languages and programming*, pages 1–12. Springer, 2006.

[Eck06] C. Eckert. *IT-Sicherheit: Konzepte, Verfahren, Protokolle*. Oldenbourg, 2006.

[ElG85] T. ElGamal. A public key cryptosystem and a signature scheme based on discrete logarithms. In G. Blakley and D. Chaum, editors, *Advances in Cryptology*, volume 196 of *Lecture Notes in Computer Science*, pages 10–18. Springer Berlin Heidelberg, 1985.

[EY07] G. Elahi and E. Yu. A goal oriented approach for modeling and analyzing security trade-offs. In C. Parent, K.-D. Schewe, V. Storey, and B. Thalheim, editors, *Conceptual Modeling - ER 2007*, volume 4801 of *Lecture Notes in Computer Science*, pages 375–390. Springer Berlin Heidelberg, 2007.

[FK00] W. Ford and J. Kaliski, B.S. Server-assisted generation of a strong secret from a password. In *Proceedings of the 9th IEEE International Workshops on Enabling Technologies: Infrastructure for Collaborative Enterprises (WET ICE)*, pages 176–180, 2000.

[For11] S. Foresti. *Preserving Privacy in Data Outsourcing*. Springer, Berlin Heidelberg, 2011.

[FWCY10] B. C. M. Fung, K. Wang, R. Chen, and P. S. Yu. Privacy-preserving data publishing: A survey of recent developments. *ACM Computing Surveys (CSUR)*, 42(4):14:1–14:53, June 2010.

[FWFP08] B. C. M. Fung, K. Wang, A. W.-C. Fu, and J. Pei. Anonymity for continuous data publishing. In *Proceedings of the International Conference on Extending Database Technology*, EDBT '08, pages 264–275, 2008.

[FWY07] B. Fung, K. Wang, and P. Yu. Anonymizing classification data for privacy preservation. *IEEE Transactions on Knowledge and Data Engineering*, 19(5):711–725, May 2007.

[GHH+14] P. Grofig, M. Härterich, I. Hang, F. Kerschbaum, M. Kohler, A. Schaad, A. Schröpfer, and W. Tighzert. Experiences and observations on the industrial implementation of a system to search over outsourced encrypted data. In *Sicherheit 2014: Sicherheit, Schutz und Zuverlässigkeit, Beiträge der 7. Jahrestagung des Fachbereichs Sicherheit der Gesellschaft für Informatik e.V. (GI), 19.-21. März 2014, Wien, Österreich*, pages 115–125, 2014.

[GMR88] S. Goldwasser, S. Micali, and R. L. Rivest. A digital signature scheme secure against adaptive chosen-message attacks. *SIAM Journal on Computing*, 17(2):281–308, 1988.

[GMR06] C. Gentry, P. MacKenzie, and Z. Ramzan. A method for making password-based key exchange resilient to server compromise. In C. Dwork, editor, *Advances in Cryptology - CRYPTO 2006*, volume 4117 of *Lecture Notes in Computer Science*, pages 142–159. Springer Berlin Heidelberg, 2006.

[GN72] R. S. Garfinkel and G. L. Nemhauser. *Integer programming*, volume 4. Wiley New York, 1972.

[GO96] O. Goldreich and R. Ostrovsky. Software protection and simulation on oblivious rams. *Journal of the ACM*, 43(3):431–473, May 1996.

[GSKK07] P. Gola, R. Schomerus, C. Klug, and B. Körffer. *BDSG: Bundesdatenschutzgesetz*. CH Beck, 2007.

[GSMB03] E.-J. Goh, H. Shacham, N. Modadugu, and D. Boneh. Sirius: Securing remote untrusted storage. In *NDSS*, volume 3, pages 131–145, 2003.

[GTF+11] V. Ganapathy, D. Thomas, T. Feder, H. Garcia-Molina, and R. Motwani. Distributing data for secure database services. In *Proceedings of the 4th International Workshop on Privacy and Anonymity in the Information Society*, PAIS '11, pages 8:1–8:10, New York, NY, USA, 2011. ACM.

[HBN11] Y. He, S. Barman, and J. Naughton. Preventing equivalence attacks in updated, anonymized data. In *Proceedings of the IEEE 27th International Conference on Data Engineering (ICDE)*, pages 529–540, 2011.

[HCH+05] J. Hughes, S. Cantor, J. Hodges, F. Hirsch, P. Mishra, R. Philpott, and E. Maler. Profiles for the OASIS security assertion markup language (SAML) v2.0, March 2005.

[HGSB13] M. Huber, M. Gabel, M. Schulze, and A. Bieber. Cumulus4j: A provably secure database abstraction layer. In A. Cuzzocrea, C. Kittl, D. Simos, E. Weippl, and L. Xu, editors, *Security Engineering and Intelligence Informatics*, volume 8128 of *Lecture Notes in Computer Science*, pages 180–193. Springer Berlin Heidelberg, 2013.

[HHK+10] C. Henrich, M. Huber, C. Kempka, J. Mueller-Quade, and R. Reussner. Technical report: Secure cloud computing through a separation of duties. *Institut für Kryptographie und Sicherheit (KIT)*, 2010.

[HILM02] H. Hacıgümüş, B. Iyer, C. Li, and S. Mehrotra. Executing SQL over encrypted data in the database-service-provider model. In *Proceedings of the ACM International Conference on Management of Data*, SIGMOD '02, pages 216–227, New York, NY, USA, 2002. ACM.

[HILM09] P. J. Haas, I. F. Ilyas, G. M. Lohman, and V. Markl. Discovering and exploiting statistical properties for query optimization in relational databases: A survey. *Statistical Analysis and Data Mining*, 1(4):223–250, 2009.

[HIM04] H. Hacıgümüş, B. Iyer, and S. Mehrotra. Efficient execution of aggregation queries over encrypted relational databases. In Y. Lee, J. Li, K.-Y. Whang, and D. Lee, editors, *Database Systems for Advanced Applications*, volume 2973 of *Lecture Notes in Computer Science*, pages 125–136. Springer, Berlin Heidelberg, 2004.

[HK99] S. Halevi and H. Krawczyk. Public-key cryptography and password protocols. *ACM Transactions on Information System Security*, 2(3):230–268, August 1999.

[HMCK12] B. Hore, S. Mehrotra, M. Canim, and M. Kantarcioglu. Secure multi-dimensional range queries over outsourced data. *The VLDB Journal*, 21(3):333–358, June 2012.

[HVO12] C. Herley and P. Van Oorschot. A research agenda acknowledging the persistence of passwords. *Security & Privacy, IEEE*, 10(1):28–36, 2012.

[HWC13] H. Hartenstein, T. Walter, and P. Castellaz. Aktuelle Umsetzungskonzepte der Universitäten des Landes Baden-Württemberg für Hochleistungsrechnen und datenintensive Dienste. *PIK-Praxis der Informationsverarbeitung und Kommunikation*, 36(2):99–108, 2013.

[Hö11] T. Höllrigl. *Informationskonsistenz im föderativen Identitätsmanagement: Modellierung und Mechanismen.* PhD thesis, Karlsruhe Institute of Technology, KIT, Karlsruhe, Germany, [in German], 2011.

[IKK12] M. Islam, M. Kuzu, and M. Kantarcioglu. Access pattern disclosure on searchable encryption: Ramification, attack and mitigation. In *Network and Distributed System Security Symposium (NDSS'12)*, 2012.

[Ioa96] Y. E. Ioannidis. Query optimization. *ACM Computing Survey (CSUR)*, 28(1):121–123, March 1996.

[ISN89] M. Ito, A. Saito, and T. Nishizeki. Secret sharing scheme realizing general access structure. *Electronics and Communications in Japan (Part III: Fundamental Electronic Science)*, 72(9):56–64, 1989.

[ISO07a] ISO standard. ISO/IEC 15939: Systems and software engineering – measurement process. *International Organization for Standarization*, 2007.

[ISO07b] ISO standard. ISO/IEC 25020: Software and system engineering–software product quality requirements and evaluation (square)–measurement reference model and guide. *International Organization for Standarization*, 2007.

[ISO07c] ISO standard. ISO/IEC 25030: Software engineering – software product quality requirements and evaluation (SQuaRE) – quality requirements. *International Organization for Standarization*, 2007.

[ISO11a] ISO standard. ISO/IEC 24760-1:2011: Information technology – Security techniques – A framework for identity management – Part 1: Terminology and concepts. *International Organization for Standardization*, 2011.

[ISO11b] ISO standard. ISO/IEC 25010: Systems and software engineering–systems and software quality requirements and evaluation (square)–system and software quality models. *International Organization for Standardization*, 2011.

[ISO11c] ISO standard. ISO/IEC 25040: Systems and software engineering - systems and software quality requirements and evaluation (SQuaRE) - evaluation process. *International Organization for Standarization*, 2011.

[ISO11d] ISO/IEC standard. ISO/IEC 27005: Information technology - security techniques - information security risk management. *International Organization for Standardization*, 2011.

[ISO13] ISO standard. ISO 27001: Information security management system (ISMS) standard. *International Organization for Standardization*, 2013.

[ISO14a] ISO standard. ISO/IEC 25000: Systems and software engineering — systems and software quality requirements and evaluation (square) — guide to square. *International Organization for Standardization*, 2014.

[ISO14b] ISO standard. ISO/IEC 27000: Information technology — security techniques — information security management systems — overview and vocabulary. *International Organization for Standardization*, 2014.

[ISO14c] ISO/IEC standard. ISO/IEC 27018: Information technology - security techniques - code of practice for protection of personally identifiable information (PII) in public clouds acting as PII processors. *International Organization for Standardization*, 2014.

[Iye02] V. S. Iyengar. Transforming data to satisfy privacy constraints. In *Proceedings of the Eighth ACM SIGKDD International Conference on Knowledge Discovery and Data Mining*, KDD '02, pages 279–288, New York, NY, USA, 2002. ACM.

[Jab96] D. P. Jablon. Strong password-only authenticated key exchange. *ACM SIGCOMM Computer Communication Review,*, 26(5):5–26, October 1996.

[Jab01] D. P. Jablon. Password authentication using multiple servers. In *Proceedings of the Conference on Topics in Cryptology: The Cryptographer's Track at RSA (CT-RSA)*, pages 344–360, 2001.

[Jai91] R. Jain. The art of computer system performance analysis: techniques for experimental design, measurement, simulation and modeling. *John Wiley*, 1991.

[JK07] A. Juels and B. S. Kaliski, Jr. Pors: Proofs of retrievability for large files. In *Proceedings of the 14th ACM Conference on Computer and Communications Security*, CCS '07, pages 584–597, New York, NY, USA, 2007. ACM.

[JKH12] K. Jünemann, J. Köhler, and H. Hartenstein. Data outsourcing simplified: Generating data connectors from confidentiality and access policies. In *Proceedings of the Workshop on Data-intensive Process Management in Large-Scale Sensor Systems (CCGrid-DPMSS)*, 2012.

[JSSSE14] J. James Stephen, S. Savvides, R. Seidel, and P. Eugster. Program analysis for secure big data processing. In *Proceedings of the 29th ACM/IEEE International Conference on Automated Software Engineering*, ASE '14, pages 277–288, New York, NY, USA, 2014. ACM.

[Jü15] K. Jünemann. *Confidential Data-Outsourcing and Self-Optimizing P2P-Networks - Coping with the Challenges of Multi-Party Systems.* PhD thesis, Karlsruhe Institute of Technology, KIT, Karlsruhe, Germany, 2015.

[Kal00] B. Kaliski. PKCS #5: Password-Based Cryptography Specification Version 2.0. RFC 2898 (Informational), September 2000.

[Kas14] KASTEL - Kompetenzzentrum für angewandte Sicherheitstechnologie - Evaluation Report. Technical report, Karlsruhe Institute of Technology, September 2014.

[KH13] J. Köhler and H. Hartenstein. Occasio: an operable concept for confidential and secure identity outsourcing. In *Proceedings of the IFIP/IEEE International Symposium on Integrated Network Management*, IM '13, 2013.

[KH14] J. Köhler and H. Hartenstein. Index optimization for l-diversified database-as-a-service. In *Proceedings of the 9th International ESORICS Workshop on Data Privacy Management*, DPM-ESORICS '14, 2014.

[KHK+13] F. Kerschbaum, M. Härterich, M. Kohler, I. Hang, A. Schaad, A. Schröpfer, and W. Tighzert. An encrypted in-memory column-store: The onion selection problem. In A. Bagchi and I. Ray, editors, *Information Systems Security*, volume 8303 of *Lecture Notes in Computer Science*, pages 14–26. Springer Berlin Heidelberg, 2013.

[KIT07] H. Koshutanski, M. Ion, and L. Telesca. Distributed identity management model for digital ecosystems. In *Proceedings of the International Conference on Emerging Security Information, Systems, and Technologies (SECURWARE)*, pages 132 –138, 2007.

[KJ14] J. Köhler and K. Jünemann. Securus: From confidentiality and access requirements to data outsourcing solutions. In M. Hansen, J.-H. Hoepman, R. Leenes, and D. Whitehouse, editors, *Privacy and Identity Management for Emerging Services and Technologies*, volume 421 of *IFIP Advances in Information and Communication Technology*, pages 139–149. Springer Berlin Heidelberg, 2014.

[KJH14] J. Köhler, K. Jünemann, and H. Hartenstein. Securus: Composition of confidentiality preserving indexing approaches for secure database-as-a-service. *PIK-Praxis der Informationsverarbeitung und Kommunikation*, 37:149–155, 2014.

[KJH15] J. Köhler, K. Jünemann, and H. Hartenstein. Confidential database-as-a-service approaches: Taxonomy and survey. *Journal of Cloud Computing*, 4(1):1–14, 2015.

[KL07] J. Katz and Y. Lindell. *Introduction to modern cryptography: principles and protocols*. CRC Press, 2007.

[KLS+12] J. Köhler, S. Labitzke, M. Simon, M. Nussbaumer, and H. Hartenstein. Facius: An easy-to-deploy saml-based approach to federate non web-based services. In *Proceedings of the 11th IEEE International Conference on Trust, Security and Privacy in Computing and Communications*, TrustCom '12, 2012.

[KLS+14] J. Köhler, S. Labitzke, M. Simon, T. Dussa, M. Nussbaumer, and H. Hartenstein. bwIDM–federated access to IT-based services at the universities of the state of Baden-Württemberg. *PIK-Praxis der Informationsverarbeitung und Kommunikation*, 37:15–21, 2014.

[KMH13] J. Köhler, J. Mittag, and H. Hartenstein. User-centric management of distributed credential repositories: Balancing availability and vulnerability. In *Proceedings of the 18th ACM Symposium on Access Control Models and Technologies*, SACMAT '13, pages 237–248, New York, NY, USA, 2013. ACM.

[KO97] E. Kushilevitz and R. Ostrovsky. Replication is not needed: single database, computationally-private information retrieval. In *Proceedings of the 38th Annual Symposium on Foundations of Computer Science*, pages 364–373, 1997.

[KPR12] S. Kamara, C. Papamanthou, and T. Roeder. Dynamic searchable symmetric encryption. In *Proceedings of the 2012 ACM Conference on Computer and Communications Security*, CCS '12, pages 965–976, New York, NY, USA, 2012. ACM.

[KSNH13] J. Köhler, M. Simon, M. Nussbaumer, and H. Hartenstein. Federating hpc access via saml: Towards a plug-and-play solution. In J. Kunkel, T. Ludwig, and H. Meuer, editors, *Supercomputing*, volume 7905 of *Lecture Notes in Computer Science*, pages 462–473. Springer Berlin Heidelberg, 2013.

[KT51] H. W. Kuhn and A. W. Tucker. Nonlinear programming, 1951.

[KV08] F. Kerschbaum and J. Vayssière. Privacy-preserving data analytics as an outsourced service. In *Proceedings of the 2008 ACM workshop on secure web services*, SWS '08, pages 87–96, New York, NY, USA, 2008. ACM.

[LD60] A. H. Land and A. G. Doig. An automatic method of solving discrete programming problems. *Econometrica: Journal of the Econometric Society*, pages 497–520, 1960.

[LDR05] K. LeFevre, D. J. DeWitt, and R. Ramakrishnan. Incognito: Efficient full-domain k-anonymity. In *Proceedings of the ACM International Conference on Management of Data*, SIGMOD '05, pages 49–60, New York, NY, USA, 2005. ACM.

[LDR06] K. LeFevre, D. J. DeWitt, and R. Ramakrishnan. Workload-aware anonymization. In *Proceedings of the 12th ACM SIGKDD International Conference on Knowledge Discovery and Data Mining*, KDD '06, pages 277–286, New York, NY, USA, 2006. ACM.

[Len83] H. W. Lenstra. Integer programming with a fixed number of variables. *Mathematics of operations research*, 8(4):538–548, 1983.

[LKDDN10] J. Lapon, M. Kohlweiss, B. De Decker, and V. Naessens. Performance analysis of accumulator based revocation mechanisms. In K. Rannenberg, V. Varadharajan, and C. Weber, editors, *Security and Privacy - Silver Linings in the Cloud*, volume 330 of *IFIP Advances in Information and Communication Technology*, pages 289–301. Springer Berlin Heidelberg, 2010.

[LLV07] N. Li, T. Li, and S. Venkatasubramanian. t-closeness: Privacy beyond k-anonymity and l-diversity. In *ICDE*, volume 7, pages 106–115, 2007.

[LO05] J. Li and E. R. Omiecinski. Efficiency and security trade-off in supporting range queries on encrypted databases. In *Proceedings of the 19th annual IFIP WG 11.3 working conference on Data and Applications Security*, DBSec'05, pages 69–83, Berlin, Heidelberg, 2005. Springer-Verlag.

[Mer79] R. C. Merkle. *Secrecy, authentication, and public key systems*. PhD thesis, Stanford University, Stanford, CA, USA, 1979.

[Mer89] R. C. Merkle. A certified digital signature. In *Proceedings on Advances in cryptology (CRYPTO)*, pages 218–238, 1989.

[MKGV07] A. Machanavajjhala, D. Kifer, J. Gehrke, and M. Venkitasubramaniam. L-diversity: Privacy beyond k-anonymity. *ACM Transactions on Knowledge Discovery from Data (TKDD)*, 1(1), March 2007.

[MLY05] S. Myagmar, A. J. Lee, and W. Yurcik. Threat modeling as a basis for security requirements. In *Symposium on requirements engineering for information security (SREIS)*, volume 2005, pages 1–8, 2005.

[MNT06] E. Mykletun, M. Narasimha, and G. Tsudik. Authentication and integrity in outsourced databases. *ACM Transactions on Storage (TOS)*, 2(2):107–138, May 2006.

[MS93] J. Moffett and M. Sloman. Policy hierarchies for distributed systems management. *Selected Areas in Communications, IEEE Journal on*, 11(9):1404–1414, Dec 1993.

[MSJ02] P. MacKenzie, T. Shrimpton, and M. Jakobsson. Threshold password-authenticated key exchange. In M. Yung, editor, *Advances in Cryptology — CRYPTO 2002*, volume 2442 of *Lecture Notes in Computer Science*, pages 385–400. Springer Berlin Heidelberg, 2002.

[MT05] E. Mykletun and G. Tsudik. Incorporating a secure coprocessor in the database-as-a-service model. In *Innovative Architecture for Future Generation High-Performance Processors and Systems*, pages 1–7, 2005.

[Nat12] National Institute of Standards and Technology. Recommendation for key management. pages 800–57 Part 1 Rev. 3, July 2012.

[NBKT11] M. Nabeel, E. Bertino, M. Kantarcioglu, and B. Thuraisingham. Towards privacy preserving access control in the cloud. In *Proceedings of the 7th International Conference on Collaborative Computing: Networking, Applications and Worksharing (CollaborateCom)*, pages 172–180, 2011.

[NC11] A. E. Nergiz and C. Clifton. Query processing in private data outsourcing using anonymization. In *Data and Applications Security and Privacy XXV*, volume 6818 of *Lecture Notes in Computer Science*, pages 138–153. Springer Berlin Heidelberg, 2011.

[NCM13] A. E. Nergiz, C. Clifton, and Q. M. Malluhi. Updating outsourced anatomized private databases. In *Proceedings of the International Conference on Extending Database Technology*, EDBT '13, pages 179–190, 2013.

[Nie06] A. E. Nieland. National security letters and the amended PATRIOT act. *Cornell L. Rev.*, 92:1201, 2006.

[NK15] T. Nilges and J. Köhler. Securus - a provably secure database outsourcing scheme. In *Proceedings of the Future Security Conference*, Berlin, September 2015. To appear.

[NM65] J. A. Nelder and R. Mead. A simplex method for function minimization. *The Computer Journal*, 7(4):308–313, 1965.

[NT06] M. Narasimha and G. Tsudik. Authentication of outsourced databases using signature aggregation and chaining. In *Proceedings of the 11th International Conference on Database Systems for Advanced Applications (DASFAA)*, pages 420–436, 2006.

[OAS13] OASIS standard. eXtensible access control markup language (XACML) version 3.0, 2013.

[OG10] F. Olumofin and I. Goldberg. Privacy-preserving queries over relational databases. In *Proceedings of the 10th international conference on Privacy enhancing technologies*, PETS'10, pages 75–92, Berlin, Heidelberg, 2010. Springer-Verlag.

[OU98] T. Okamoto and S. Uchiyama. A new public-key cryptosystem as secure as factoring. In K. Nyberg, editor, *Advances in Cryptology — EUROCRYPT'98*, volume 1403 of *Lecture Notes in Computer Science*, pages 308–318. Springer, Berlin Heidelberg, 1998.

[Pai99] P. Paillier. Public-key cryptosystems based on composite degree residuosity classes. In *Proceedings of EUROCRYPT*, 1999.

[Pap81] C. H. Papadimitriou. On the complexity of integer programming. *Journal of the ACM*, 28(4):765–768, October 1981.

[PGK88] D. A. Patterson, G. Gibson, and R. H. Katz. A case for redundant arrays of inexpensive disks (RAID). In *Proceedings of the ACM International Conference on Management of Data*, SIGMOD '88, pages 109–116, 1988.

[PHB06] A. Pretschner, M. Hilty, and D. Basin. Distributed usage control. *Communications of the ACM*, 49(9):39–44, 2006.

[PLZ13] R. Popa, F. Li, and N. Zeldovich. An ideal-security protocol for order-preserving encoding. In *Proceedings of the IEEE Symposium on Security and Privacy*, S&P '13, pages 463–477, May 2013.

[PR91] M. Padberg and G. Rinaldi. A branch-and-cut algorithm for the resolution of large-scale symmetric traveling salesman problems. *SIAM review*, 33(1):60–100, 1991.

[PRZB11] R. A. Popa, C. M. S. Redfield, N. Zeldovich, and H. Balakrishnan. Cryptdb: Protecting confidentiality with encrypted query processing. In *Proceedings of the Twenty-Third ACM Symposium on Operating Systems Principles*, SOSP '11, pages 85–100, New York, NY, USA, 2011. ACM.

[PXW⁺07] J. Pei, J. Xu, Z. Wang, W. Wang, and K. Wang. Maintaining k-anonymity against incremental updates. In *Scientific and Statistical Database Management, 2007. SSBDM '07. 19th International Conference on*, pages 5–5, July 2007.

[PZB11] R. Popa, N. Zeldovich, and H. Balakrishnan. Cryptdb: A practical encrypted relational DBMS. Technical report, MIT, 2011.

[PZM13] H. Pang, J. Zhang, and K. Mouratidis. Enhancing access privacy of range retrievals over B+-trees. *Knowledge and Data Engineering, IEEE Transactions on*, 25(7):1533–1547, 2013.

[RAD78] R. L. Rivest, L. Adleman, and M. L. Dertouzos. On data banks and privacy homomorphisms. *Foundations of secure computation*, 4(11):169–180, 1978.

[RBO⁺10] R. Ranchal, B. Bhargava, L. Othmane, L. Lilien, A. Kim, M. Kang, and M. Linderman. Protection of identity information in cloud computing without trusted third party. In *Proceedings of the 29th IEEE Symposium on Reliable Distributed Systems*, pages 368–372, 2010.

[Rei84] S. P. Reiss. Practical data-swapping: The first steps. *ACM Transactions on Database Systems*, 9(1):20–37, March 1984.

[RSN12] S. Ruj, M. Stojmenovic, and A. Nayak. Privacy preserving access control with authentication for securing data in clouds. In *Proceedings of the 12th IEEE/ACM International Symposium on Cluster, Cloud and Grid Computing (CCGrid)*, pages 556 –563, 2012.

[RWRS00] C. Rigney, S. Willens, A. Rubens, and W. Simpson. Remote authentication dial in user service (radius) - RfC 2865. *IETF*, 2000.

[Sam01] P. Samarati. Protecting respondents identities in microdata release. *IEEE Transactions on Knowledge and Data Engineering*, 13(6):1010–1027, Nov 2001.

[San03] R. Sandhu. Good-enough security: Toward a pragmatic business-driven discipline. *IEEE Internet Computing*, 7(1):66–68, 2003.

[SBG02] R. Sandhu, M. Bellare, and R. Ganesan. Password-enabled PKI: Virtual smartcards versus virtual soft tokens. In *Proceedings of the Annual PKI Research Workshop*, 2002.

[SC07] R. Sion and B. Carbunar. On the computational practicality of private information retrieval. In *Proceedings of the Network and Distributed Systems Security Symposium*, 2007.

[Sch11] B. Schneier. *Secrets and lies: digital security in a networked world.* John Wiley & Sons, 2011.

[Sda14] C. Sdannowitz. Die eigenen Mitarbeiter als Datendiebe - Insider-Threats verhindern. *kes special - Die Zeitschrift für Informations-Sicherheit*, 30(4):22–23, August 2014.

[Sha79] A. Shamir. How to share a secret. *Communications of the ACM*, 22(11):612–613, November 1979.

[SKR12] A. Sabouri, I. Krontiris, and K. Rannenberg. Attribute-based credentials for trust (abc4trust). In S. Fischer-Hübner, S. Katsikas, and G. Quirchmayr, editors, *Trust, Privacy and Security in Digital Business*, volume 7449 of *Lecture Notes in Computer Science*, pages 218–219. Springer Berlin Heidelberg, 2012.

[Slo94] M. Sloman. Policy driven management for distributed systems. *Journal of Network and Systems Management*, 2(4):333–360, 1994.

[SLX01] G. Stone, B. Lundy, and G. Xie. Network policy languages: a survey and a new approach. *Network, IEEE*, 15(1):10–21, Jan 2001.

[SS13] E. Stefanov and E. Shi. Multi-cloud oblivious storage. In *Proceedings of the 2013 ACM SIGSAC Conference on Computer and Communications Security*, CCS '13, pages 247–258, New York, NY, USA, 2013. ACM.

[SSMJF12] M. Stihler, A. O. Santin, A. L. Marcon Jr., and J. d. S. Fraga. Integral federated identity management for cloud computing. In *Proceedings of the 5th International Conference on New Technologies, Mobility and Security (NTMS)*, pages 1 –5, 2012.

[SSSE14] J. J. Stephen, S. Savvides, R. Seidel, and P. Eugster. Practical confidentiality preserving big data analysis. In *Proceedings of the 6th USENIX Conference on Hot Topics in Cloud Computing*, HotCloud'14, pages 10–10, Berkeley, CA, USA, 2014. USENIX Association.

[STW95] M. Steiner, G. Tsudik, and M. Waidner. Refinement and extension of encrypted key exchange. *ACM SIGOPS Operating Systems Review*, 29(3):22–30, July 1995.

[SvDS+13] E. Stefanov, M. van Dijk, E. Shi, C. Fletcher, L. Ren, X. Yu, and S. Devadas. Path ORAM: An extremely simple oblivious RAM protocol. In *Proceedings of the ACM SIGSAC Conference on Computer and Communications Security*, CCS '13, pages 299–310, New York, NY, USA, 2013. ACM.

[SW08] H. Shacham and B. Waters. Compact proofs of retrievability. In J. Pieprzyk, editor, *Advances in Cryptology - ASIACRYPT 2008*, volume 5350 of *Lecture Notes in Computer Science*, pages 90–107. Springer Berlin Heidelberg, 2008.

[Swe02a] L. Sweeney. Achieving k-anonymity privacy protection using generalization and suppression. *International Journal of Uncertainty, Fuzziness and Knowledge-Based Systems*, 10(05):571–588, 2002.

[Swe02b] L. Sweeney. k-anonymity: A model for protecting privacy. *International Journal of Uncertainty, Fuzziness and Knowledge-Based Systems*, 10(05):557–570, 2002.

[SWP00] D. X. Song, D. Wagner, and A. Perrig. Practical techniques for searches on encrypted data. In *Proceedings of the IEEE Symposium on Security and Privacy*, S&P '00, pages 44–55, 2000.

[TBF14] M. Thoss, W. Bachmann, and J. Flemming. Perpektivwechsel - Drei Kritische Blickwinkel zum Datenschutz im Krankenhaus. *kes - Die Zeitschrift für Informations-Sicherheit*, 30(4):54–60, Aug. 2014.

[TLMM13] S. D. Tetali, M. Lesani, R. Majumdar, and T. Millstein. MrCrypt: Static analysis for secure cloud computations. In *Proceedings of the ACM SIGPLAN International Conference on Object Oriented Programming Systems Languages and Applications*, OOPSLA '13, pages 271–286, New York, NY, USA, 2013. ACM.

[Tra10] Transaction Processing Performance Council. TPC-C standard specification - revision 5.11. *Published at http://www.tpc.org/*, 2010.

[Tra14] Transaction Processing Performance Council. TPC-H standard specification - revision 2.17.1. *Published at http://www.tpc.org/*, 2014.

[Tsa93] E. Tsang. *Foundations of constraint satisfaction*, volume 289. Academic press London, 1993.

[VCDC11] P. Victor, C. Cornelis, and M. De Cock. *Trust networks for recommender systems*, volume 4. Springer Science & Business Media, 2011.

[vD95] M. van Dijk. A linear construction of perfect secret sharing schemes. In A. De Santis, editor, *Advances in Cryptology — EUROCRYPT'94*, volume 950 of *Lecture Notes in Computer Science*, pages 23–34. Springer Berlin Heidelberg, 1995.

[WACS10] M. Weir, S. Aggarwal, M. Collins, and H. Stern. Testing metrics for password creation policies by attacking large sets of revealed passwords. In *Proceedings of the 17th ACM Conference on Computer and Communications Security*, CCS '10, pages 162–175, New York, NY, USA, 2010. ACM.

[WFY07] K. Wang, B. Fung, and P. Yu. Handicapping attacker's confidence: an alternative to k-anonymization. *Knowledge and Information Systems*, 11(3):345–368, 2007.

[WY14] W. Wei and T. Yu. Integrity assurance for outsourced databases without DBMS modification. In V. Atluri and G. Pernul, editors, *Data and Applications Security and Privacy XXVIII*, volume 8566 of *Lecture Notes in Computer Science*, pages 1–16. Springer Berlin Heidelberg, 2014.

[WYX13] W. Wei, T. Yu, and R. Xue. iBigTable: Practical data integrity for Bigtable in public cloud. In *Proceedings of the Third ACM Conference on Data and Application Security and Privacy*, CODASPY '13, pages 341–352, New York, NY, USA, 2013. ACM.

[XT06] X. Xiao and Y. Tao. Anatomy: Simple and effective privacy preservation. In *Proceedings of the 32Nd International Conference on Very Large Data Bases*, VLDB '06, pages 139–150. VLDB Endowment, 2006.

[XT07] X. Xiao and Y. Tao. M-invariance: towards privacy preserving republication of dynamic datasets. In *Proceedings of the ACM International Conference on Management of Data*, SIGMOD '07, pages 689–700, 2007.

[XWYM07] M. Xie, H. Wang, J. Yin, and X. Meng. Integrity auditing of outsourced data. In *Proceedings of the 33rd International Conference on Very large data bases*, pages 782–793, 2007.

[XWYM08] M. Xie, H. Wang, J. Yin, and X. Meng. Providing freshness guarantees for outsourced databases. In *Proceedings of the 11th international conference on Extending database technology: Advances in database technology*, EDBT '08, pages 323–332, 2008.

[YY05] F. Yao and Y. Yin. Design and analysis of password-based key derivation functions. In A. Menezes, editor, *Topics in Cryptology - CT-RSA 2005*, volume 3376 of *Lecture Notes in Computer Science*, pages 245–261. Springer Berlin Heidelberg, 2005.

[ZHP+09] B. Zhou, Y. Han, J. Pei, B. Jiang, Y. Tao, and Y. Jia. Continuous privacy preserving publishing of data streams. In *Proceedings of the International Conference on Extending Database Technology*, EDBT '09, pages 648–659, 2009.